tinyAVR®
Microcontroller
Projects for
the Evil Genius™

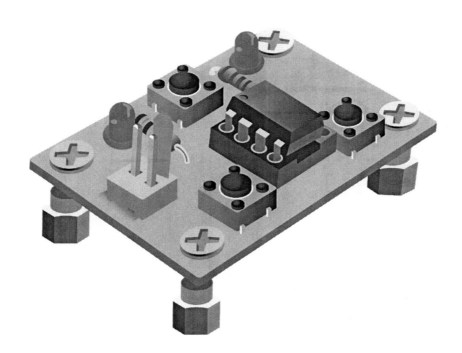

Evil Genius™ Series

Bike, Scooter, and Chopper Projects for the Evil Genius

Bionics for the Evil Genius: 25 Build-It-Yourself Projects

Electronic Circuits for the Evil Genius, Second Edition: 64 Lessons with Projects

Electronic Gadgets for the Evil Genius: 28 Build-It-Yourself Projects

Electronic Sensors for the Evil Genius: 54 Electrifying Projects

50 Awesome Auto Projects for the Evil Genius

50 Green Projects for the Evil Genius

50 Model Rocket Projects for the Evil Genius

51 High-Tech Practical Jokes for the Evil Genius

46 Science Fair Projects for the Evil Genius

Fuel Cell Projects for the Evil Genius

Holography Projects for the Evil Genius

Mechatronics for the Evil Genius: 25 Build-It-Yourself Projects

Mind Performance Projects for the Evil Genius: 19 Brain-Bending Bio Hacks

MORE Electronic Gadgets for the Evil Genius: 40 NEW Build-It-Yourself Projects

101 Outer Space Projects for the Evil Genius

101 Spy Gadgets for the Evil Genius

125 Physics Projects for the Evil Genius

123 PIC® Microcontroller Experiments for the Evil Genius

123 Robotics Experiments for the Evil Genius

PC Mods for the Evil Genius: 25 Custom Builds to Turbocharge Your Computer

PICAXE Microcontroller Projects for the Evil Genius

Programming Video Games for the Evil Genius

Recycling Projects for the Evil Genius

Solar Energy Projects for the Evil Genius

Telephone Projects for the Evil Genius

30 Arduino Projects for the Evil Genius

25 Home Automation Projects for the Evil Genius

22 Radio and Receiver Projects for the Evil Genius

tinyAVR® Microcontroller Projects for the Evil Genius™

Dhananjay V. Gadre and Nehul Malhotra

New York   Chicago   San Francisco   Lisbon   London   Madrid
Mexico City   Milan   New Delhi   San Juan   Seoul
Singapore   Sydney   Toronto

The McGraw·Hill Companies

Cataloging-in-Publication Data is on file with the Library of Congress

McGraw-Hill books are available at special quantity discounts to use as premiums and sales promotions, or for use in corporate training programs. To contact a representative, please e-mail us at bulksales@ mcgraw-hill.com.

tinyAVR® Microcontroller Projects for the Evil Genius™

1 2 3 4 5 6 7 8 9 0   QDB QDB   1 0 9 8 7 6 5 4 3 2 1

ISBN 978-0-07-174454-6
MHID   0-07-174454-1

Sponsoring Editor Roger Stewart	**Indexer** Claire Splan
Editorial Supervisor Janet Walden	**Production Supervisor** Jean Bodeaux
Acquisitions Coordinator Joya Anthony	**Composition** TypeWriting
Project Manager Patricia Wallenburg	**Art Director, Cover** Jeff Weeks
Copy Editor Lisa McCoy	**Cover Designer** Todd Radom
Proofreader Paul Tyler	

This book is dedicated to Professor Shailaja M. Karandikar (1920–1995), in whose spacious home with a mini library I was always welcome to browse and borrow any book.

And to Professor Neil Gershenfeld, who made it possible to write this one!

—Dhananjay V. Gadre

To my parents, who have given me my identity. And to my sister, Neha, who is my identity!

—Nehul Malhotra

About the Authors

Dhananjay V. Gadre (New Delhi, India) completed his MSc (electronic science) from the University of Delhi and MEng (computer engineering) from the University of Idaho. In his professional career of more than 21 years, he has taught at the SGTB Khalsa College, University of Delhi, worked as a scientific officer at the Inter University Centre for Astronomy and Astrophysics (IUCAA), Pune, and since 2001, has been with the Electronics and Communication Engineering Division, Netaji Subhas Institute of Technology, New Delhi, currently as an associate professor. He is also associated with the global Fablab network and is a faculty member at the Fab Academy. Professor Gadre is the author of several professional articles and three books. One of his books has been translated into Chinese and another into Greek. He is a licensed radio amateur with the call sign VU2NOX and hopes to design and build an amateur radio satellite someday.

Nehul Malhotra (New Delhi, India) completed his undergraduate degree in electronics and communication engineering from the Netaji Subhas Institute of Technology, New Delhi. He worked in Professor Gadre's laboratory, collaborating extensively in the ongoing projects. He was also the founder CEO of a startup called LearnMicros. Nehul once freed a genie from a bottle he found on a beach. As a reward, he has been granted 30 hours in a day. Currently, Nehul is a graduate student at the Indian Institute of Management, Ahmedabad, India.

Contents at a Glance

1 Tour de Tiny.   1

2 LED Projects 29

3 Advanced LED Projects 55

4 Graphics LCD Projects. 99

5 Sensor Projects. 129

6 Audio Projects. 169

7 Alternate Energy Projects 191

A C Programming for AVR Microcontrollers. 213

B Designing and Fabricating PCBs 225

C Illuminated LED Eye Loupe 239

Index 247

Contents

Acknowledgments xiii

Introduction xv

1 Tour de Tiny **1**

About the Book 1
Atmel's tinyAVR Microcontrollers 2
tinyAVR Devices 2
tinyAVR Architecture 3
Elements of a Project 8
Power Sources 11
Hardware Development Tools 17
Software Development 20
Making Your Own PCB 24
Project 1 Hello World! of Microcontrollers 26
Conclusion 28

2 LED Projects **29**

LEDs 29
Types of LEDs 31
Controlling LEDs 32
Project 2 Flickering LED Candle 35
Project 3 RGB LED Color Mixer 41
Project 4 Random Color and Music Generator 45
Project 5 LED Pen 49
Conclusion 54

3 Advanced LED Projects **55**

Multiplexing LEDs 55
Charlieplexing 65
Project 6 Mood Lamp 67
Project 7 VU Meter with 20 LEDs 72
Project 8 Voltmeter 76
Project 9 Celsius and Fahrenheit Thermometer 80
Project 10 Autoranging Frequency Counter 82
Project 11 Geek Clock 84
Project 12 RGB Dice 90
Project 13 RGB Tic-Tac-Toe 93
Conclusion 97

4 Graphics LCD Projects **99**
 Principle of Operation 99
 Nokia 3310 GLCD. 101
 Project 14 Temperature Plotter 105
 Project 15 Tengu on Graphics Display 109
 Project 16 Game of Life 113
 Project 17 Tic-Tac-Toe. 117
 Project 18 Zany Clock 119
 Project 19 Rise and Shine Bell 123
 Conclusion. 128

5 Sensor Projects **129**
 LED as a Sensor 129
 Thermistor 130
 LDR 130
 Inductor as Magnetic Field Sensor 131
 Project 20 LED as a Sensor and Indicator 131
 Project 21 Valentine's Heart LED Display with Proximity Sensor 136
 Project 22 Electronic Fire-free Matchstick 140
 Project 23 Spinning LED Top with Message Display. 144
 Project 24 Contactless Tachometer 149
 Project 25 Inductive Loop-based Car Detector and Counter 153
 Project 26 Electronic Birthday Blowout Candles 159
 Project 27 Fridge Alarm 164
 Conclusion. 168

6 Audio Projects **169**
 Project 28 Tone Player. 171
 Project 29 Fridge Alarm Redux. 176
 Project 30 RTTTL Player 178
 Project 31 Musical Toy 185
 Conclusion. 189

7 Alternate Energy Projects **191**
 Choosing the Right Voltage Regulator 192
 Building the Faraday Generator 194
 Experimental Results and Discussion 195
 Project 32 Batteryless Infrared Remote. 196
 Project 33 Batteryless Electronic Dice 201
 Project 34 Batteryless Persistence-of-Vision Toy 206
 Conclusion. 212

A C Programming for AVR Microcontrollers **213**
 Differences Between ANSI C and Embedded C. 214
 Data Types and Operators 214
 Efficient Management of I/O Ports 217
 A Few Important Header Files 220
 Functions 220

Interrupt Handling 221
Arrays. 222
More C Utilities. 222

B   Designing and Fabricating PCBs. 225
EAGLE Light Edition 225
EAGLE Windows 225
EAGLE Tutorial. 226
Adding New Libraries 227
Placing the Components and Routing 228
Roland Modela MDX-20 PCB Milling Machine 228

C   Illuminated LED Eye Loupe 239
Version 2 of the Illuminated LED Eye Loupe. 242
Version 3 of the Illuminated LED Eye Loupe. 244

Index 247

Acknowledgments

WE STARTED BUILDING PROJECTS with tinyAVR microcontrollers several years ago. Designing projects using feature-constrained microcontrollers was a thrill. Slowly, the number of projects kept piling up, and we thought of documenting them with the idea of sharing them with others. The result is this book.

Many students helped with the development of the projects described in this book. They are Anurag Chugh, Saurabh Gupta, Gaurav Minocha, Mayank Jain, Harshit Jain, Hashim Khan, Nipun Jindal, Prateek Gupta, Nikhil Kautilya, Kritika Garg, and Lalit Kumar. As always, Satya Prakash at the Centre for Electronics Design and Technology (CEDT) at NSIT was a great help in fabricating many of the projects.

Initially, the project circuit boards were made on a general-purpose circuit board, or custom circuit boards were ordered through PCB manufacturers. Since 2008, when Neil Gershenfeld, professor at the Center for Bits and Atoms, Media Labs, Massachusetts Institute of Technology, presented me with a MDX20 milling machine, the speed and ease of in-house PCB fabrication increased significantly. With the MDX20 milling machine, we are able to prototype a circuit in a few hours in contrast to our previous pace of one circuit a week. The generous help of Neil Gershenfeld and his many suggestions is gratefully acknowledged. Thanks are also due to Sherry Lassiter, program manager, Center for Bits and Atoms, for supporting our activities.

Lars Thore Aarrestaad, Marco Martin Joaquim, and Imran Shariff from Atmel helped with device samples and tools.

I thank Roger Stewart, editorial director at McGraw-Hill, for having great faith in the idea of this book and Joya Anthony, acquisitions coordinator, for being persuasive but gentle even when all the deadlines were missed. Vaishnavi Sundararajan did a great job of editing the manuscript at our end before we shipped each chapter to the editors. Thank you, guys!

Nehul Malhotra, a student collaborating in several of the projects, made significant contributions to become a co-author. His persistence and ability to work hard and long hours are worth emulating by fellow students.

This book would not have been possible without Sangeeta and Chaitanya, who are my family and the most important people in my life. Thank you for your patience and perseverance!

Introduction

MORE THAN TEN YEARS AGO, when I wrote a book on AVR microcontrollers, AVRs were the new kids on the block and not many people had heard of these chips. I had to try out these new devices since I was sick of using 8051 microcontrollers, which did not offer enough features for complex requirements. Even though AVRs were new, the software tools offered by Atmel were quite robust, and I could read all about these chips and program my first application in a matter of days. Since these devices had just debuted, high-level language tools were not easily available, or were too buggy, or produced too voluminous a code even for simple programs. Thus, all the projects in that AVR book were programmed in assembly language. However, things are quite different now. The AVR microcontroller family has stabilized and currently is the second-largest-selling eight-bit microcontroller family in the whole world! Plenty of quality C compilers are available, too, for the AVR family. AVR is also supported by GCC (GNU C Compiler) as AVRGCC, which means one need not spend any money for the C compiler when choosing to use AVRGCC.

When I started using the AVR more than ten years ago, several eight-pin devices caught my attention. Up to that point, an eight-pin integrated circuit meant a 741 op-amp or 555 timer chip. But here was a complete computer in an eight-pin package. It was fascinating to see such small computers, and even more fascinating to design with them. The fascination has continued over the years. Also, Atmel wasn't sitting still with its small microcontroller series. It expanded the series and gave it a new name, tinyAVR microcontrollers, and added many devices, ranging from a six-pin part to a 28-pin device. These devices are low-cost offerings and, in volume, cost as little as 25 cents each.

Today, microcontrollers are everywhere, from TV remotes to microwave ovens to mobile phones. For the purpose of learning how to program and use these devices, people have created a variety of learning tools and kits and environments. One such popular environment is the Arduino. Arduino is based on the AVR family of microcontrollers, and instead of having to learn an assembly language or C to program, Arduino has its own language that is easy to learn—one can start using an Arduino device in a single day. It is promoted as a "low learning threshold" microcontroller system. The simplest and smallest Arduino platform uses a 28-pin AVR, the ATMega8 microcontroller, and costs upwards of $12. However, if you want to control a few LEDs or need just a couple of I/O pins for your project, you might wonder why you need a 28-pin device. Welcome to the world of tinyAVR microcontrollers!

This book illustrates 34 complete, working projects. All of these projects have been implemented with the tinyAVR series of microcontrollers and are arranged in seven chapters. The first chapter is a whirlwind tour of the AVR, specifically, the tinyAVR microcontroller architecture, the elements of a microcontroller-based project, power supply considerations, etc. The 34 projects span six themes covering LED projects, advanced LED projects, graphics LCD projects, sensor-based projects, audio projects, and finally alternative energy–powered projects. Some of these projects have already become popular and are available as products. Since all the details of these projects are described in this book, these projects make great sources of ideas for hackers and DIY enthusiasts to play with. The ideas presented in these projects can, of course, be used and improved upon. The schematic diagrams and board files for all of the projects are available and can be used to order PCBs from PCB manufacturers. Most of the components can be ordered through Digikey or Farnell.

The project files such as schematic and board files for all the projects, videos, and photographs are available on our website: www.avrgenius.com/tinyavr1.

Chapter 1: Tour de Tiny

- tinyAVR architecture, important features of tinyAVR microcontrollers, designing with microcontrollers, designing a power supply for portable applications
- Tools required for building projects, making circuit boards, the Hello World! of microcontrollers

Chapter 2: LED Projects

- Types of LEDs, their characteristics, controlling LEDs
- Four projects: LED candle, RGB LED color mixer, random color and music generator, LED pen

Chapter 3: Advanced LED Projects

- Controlling a large number of LEDs using various multiplexing techniques
- Eight projects: mood lamp, VU meter with 20-LED display, voltmeter, autoranging frequency counter, Celsius and Fahrenheit thermometer, geek clock, RGB dice, RGB tic-tac-toe

Chapter 4: Graphics LCD Projects

- Operation of LCD displays, types of LCDs, Nokia 3310 graphics LCD
- Six projects: temperature plotter, Tengu on graphics display, Game of Life, tic-tac-toe, zany clock, school bell

Chapter 5: Sensor Projects

- Various types of sensors for light, temperature, magnetic field, etc., and their operation
- Eight projects: LED as a sensor and indicator, Valentine's LED heart display with proximity sensor, electronic fire-free matchstick, spinning LED top with message display, contactless tachometer, inductive loop-based car detector and counter, electronic birthday blowout candles, fridge alarm

Chapter 6: Audio Projects

- Generating music and sound using a microcontroller
- Four projects: tone player, fridge alarm revisited, RTTTL player, musical toy

Chapter 7: Alternate Energy Projects

- Generating voltage using Faraday's law and using it to power portable applications
- Three projects: batteryless TV remote, batteryless electronic dice, batteryless POV toy

Appendix A: C Programming for AVR Microcontrollers

- A jump-start that enables readers to quickly adapt to C commands used in embedded applications and to use C to program the tinyAVR microcontrollers

Appendix B: Designing and Fabricating PCBs

■ EAGLE schematic capture and board routing program. All of the PCBs in the projects in this book are made using the free version of EAGLE. The boards can be made from PCB vendors or using the Modela (or another) PCB milling machine. Alternative construction methods also are discussed.

Appendix C: Illuminated LED Eye Loupe

■ Building a cool microcontroller-based LED eye loupe

We hope you have as much fun building these projects as we have enjoyed sharing them with you.

Tour de Tiny

THANKS TO MOORE'S LAW, silicon capacity is still doubling (well, almost) every 18 months. What that means is that after every year and a half, semiconductor integrated circuits (IC) manufacturers can squeeze in twice the number of transistors and other components in the same area of silicon. This important hypothesis was first laid down by Gordon Moore, the co-founder of Intel, in the mid-1960s, and surprisingly, it still holds true—more or less. The size of the desktop personal computers (PC) has been shrinking. From desktops to slim PCs, to cubes and handheld PCs, we have them all. Lately, another form of even smaller computers has been making the rounds: small form factor (SFF) PCs. The SFF concept shows the availability of small, general-purpose computer systems available to individual consumers, and these need not be specialized embedded systems running custom software. The impact of Moore's law is felt not only on the size of personal computers, but also on the everyday electronic devices we use; my current mobile phone, which offers me many more features than my previous one, is much smaller than its predecessor!

When we use the term "computer," it most often means the regular computing device we use to perform word processing, web browsing, etc. But almost every electronic device these days is equipped with some computing capabilities inside. Such computers are called embedded computers, since they are "embedded" inside a larger device,

making that device smarter and more capable than it would have been without this "computer."

In our quest for even smaller and sleeker computer systems and electronic gadgets, we draw our attention towards computers with an even smaller footprint: the Tiny form factor computers. Unlike the rest, these are specialized computer systems, small enough to fit in a shirt pocket. Many manufacturers provide the bare bones of such computers, and Microchip and Atmel are front-runners. With footprints as small as those of six-pin devices, not bigger than a grain of rice, all they need is a suitable power source and interface circuit. Throw in the custom software, and you have your own personal small gadget that can be as unique as you want it to be.

What can such small embedded computers do? Can they be of any use at all? We show how small they can be and what all they can do.

About the Book

The book has six project chapters. The projects in each chapter are arranged around a particular theme, such as light-emitting diodes (LEDs) or sensors. There is no particular sequence to these chapters, and they can be read in random order. If you are, however, a beginner, then it is recommended that you follow the chapters sequentially. Chapter 1 has introductory information about the project development process, tools, power supply sources, etc., and it is highly recommended even if you are an advanced reader, so that you can easily follow the style and development process that we employ in later chapters.

Atmel's tinyAVR Microcontrollers

The tinyAVR series of microcontrollers comes in many flavors now. The number of input/output (I/O) pins ranges from 4 in the smallest series, ATtiny4/5/9/10, to 28 in ATtiny48/88. Some packages of ATtiny48/88 series have 24 I/O pins only. A widely used device is ATtiny13, which has a total of eight pins, with two mandatory pins for power supply, leaving you with six I/O pins. That doesn't sound like much, but it turns out that a lot can be done even with these six I/O pins, even without having to use additional I/O expansion circuits.

From the table of tinyAVR devices presented later in this chapter, we have selected ATtiny13, ATtiny25/45/85, and ATtiny261/461/861 for most of the projects. They represent the entire spectrum of Tiny devices. All of these devices have an on-chip static random access memory (SRAM), an important requisite for programming these chips using C. Tiny13 has just 1K of program memory, while Tiny861 and Tiny85 have 8K. Tiny13 and Tiny25/45/85 are pin-compatible, but the devices of latter series have more memory and features. Whenever the code doesn't fit in Tiny13, it can be replaced with Tiny25/45/85, depending on memory requirements.

The projects that are planned for this book have a distinguishing feature: Almost all of them have fascinating visual appeal in the form of large LED-based displays. A new technique of interfacing a large number of LEDs using a relatively small number of I/O pins, called

Charlieplexing, makes it possible to interface up to 20 LEDs using just five I/O pins. This technique has been used to create appealing graphical displays or to add a seven-segment type of readout to the projects. Other projects that do not have LED displays feature graphical LCDs.

Each project can be built over a weekend and can be used gainfully in the form of a toy or an instrument.

tinyAVR Devices

tinyAVR devices vary from one another in several ways, such as the number of I/O pins, memory sizes, package type like dual in-line package (DIP), small outline integrated circuit (SOIC) or micro lead frame (MLF), peripheral features, communication interfaces, etc. Figure 1-1 shows some tinyAVRs in DIP packaging, while Figure 1-2 shows some tinyAVRs in surface mount device (SMD) SOIC packaging. The complete list

Figure 1-1 tinyAVR microcontrollers in DIP packaging

TABLE 1-1	Some Major Series/Devices of the tinyAVR Family	
S. No.	**Series/Device**	**Features**
1	ATtiny4/5/9/10	Maximum 4 I/O pins, 1.8–5.5V operation, 32B SRAM, up to 12 MIPS throughput at 12 MHz, Flash program memory 1KB in ATtiny9/10 and 512B in ATtiny4/5, analog to digital converter (ADC) present in ATtiny5/10
2	ATtiny13	Maximum 6 I/O pins, 1.8–5.5V operation, 64B SRAM, 64B EEPROM, up to 20 MIPS throughput at 20 MHz, 1KB Flash program memory, ADC
3	ATtiny24/44/84	Maximum 12 I/O pins, 1.8–5.5V operation, 128/256/512B SRAM and 128/256/512B EEPROM in ATtiny24/44/84, respectively, up to 20 MIPS throughput at 20 MHz, Flash program memory 2KB in ATtiny24, 4KB in ATtiny44, and 8KB in ATtiny84, ADC, on-chip temperature sensor, universal serial interface (USI)
4	ATtiny25/45/85	Maximum 6 I/O pins, 1.8–5.5V operation, 128/256/512B SRAM and 128/256/512B EEPROM in ATtiny25/45/85, respectively, up to 20 MIPS throughput at 20 MHz, Flash program memory 2KB in ATtiny25, 4KB in ATtiny45, and 8KB in ATtiny85, ADC, USI
5	ATtiny261/461/861	Maximum 16 I/O pins, 1.8–5.5V operation, 128/256/512B SRAM and 128/256/512B EEPROM in ATtiny261/461/861, respectively, up to 20 MIPS throughput at 20 MHz, Flash program memory 2KB in ATtiny261, 4KB in ATtiny461, and 8KB in ATtiny861, ADC, USI
6	ATtiny48/88	Maximum 24/28 I/O pins (depending upon package), 1.8–5.5V operation, 256/512B SRAM in ATtiny48/88, respectively, 64B EEPROM, up to 12 MIPS throughput at 12 MHz, Flash program memory 4KB in ATtiny48 and 8KB in ATtiny88, ADC, serial peripheral interface (SPI)
7	ATtiny43U	Maximum 16 I/O pins, 0.7–1.8V operation, 256B SRAM, 64B EEPROM, up to 1 MIPS throughput per MHz, 4KB Flash program memory, ADC, on-chip temperature sensor, USI, ultra low voltage device, integrated boost converter automatically generates a stable 3V supply voltage from a low voltage battery input down to 0.7V

of these devices is highly dynamic, as Atmel keeps adding newer devices to replace the older ones regularly. The latest changes can always be tracked on www.avrgenius.com/tinyavr1.

Most of these devices are organized in such a way that each member of the series varies from the others only in a few features, like memory size, etc. Some major series and devices of the tinyAVR family that are the main focus of this book have been summarized in Table 1-1, and are shown in Figures 1-1 and 1-2.

If you see the datasheet of any device and find that its name is suffixed by "A," it implies that it belongs to the picoPower technology AVR microcontroller class and incorporates features to reduce the power consumption on the go.

tinyAVR Architecture

This section deals with the internal details of the Tiny devices. It may be noted that this section follows a generic approach to summarize the common features of the Tiny series. Certain

Figure 1-2 tinyAVR microcontrollers in SMD packaging

features may be missing from some devices, while some additional ones may be present. For more information on these features, refer to the datasheet of the individual devices.

Memory

The AVR architecture has two main memory spaces: the data memory and the program memory space. In addition, these devices feature an electrically erasable programmable read-only memory (EEPROM) memory for data storage. The Flash program memory is organized as a linear array of 16-bit-wide locations because all the AVR instructions are either 16 bits or 32 bits wide. The internal memory SRAM uses the same address space as that used by register file and I/O registers. The lowermost 32 addresses are taken by registers, the next 64 locations are taken by I/O registers, and then the SRAM addressing continues from location 0x60. The internal EEPROM is used for

temporary nonvolatile data storage. The following illustration shows the memory map of Tiny controllers.

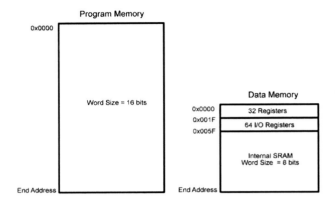

I/O Ports

Input/Output (I/O) ports of AVR devices are comprised of individual I/O pins, which can be configured individually for either input or output. Apart from this, when the pin is declared as an input, there is an option to enable or disable the pull-up on it. Enabling the pull-up is necessary to read the sensors that don't give an electrical signal, like microswitches. Each output buffer has a sink and source capability of 40mA. So, the pin driver is strong enough to drive LED displays directly. All I/O pins also have protection diodes to both VCC and Ground. The following illustration shows the block diagram of the AVR I/O ports.

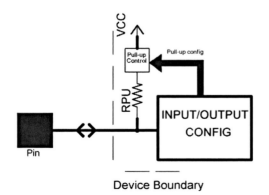

Timers

tinyAVR devices generally have eight-bit timers that can be clocked either synchronously or asynchronously. The synchronous clock sources include the device clock or its factors (the clock divided by a suitable prescaler), whereas asynchronous clock sources include the external clock or phase lock loop (PLL) clock, which goes up to 64 MHz. Some devices also include 10-bit or 16-bit timers. Besides counting, these timers also have compare units, which generate pulse width modulation on I/O pins. These timers can be run in various modes, like normal mode, capture mode, pulse width modulation (pwm) mode, clear timer on compare match, etc. Each timer has several interrupt sources associated with it, which are described in the next section on interrupts. The following illustration shows the block diagram of the AVR timer.

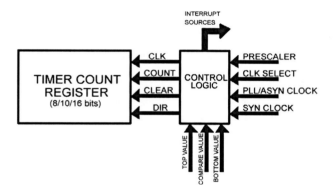

Interrupts

The AVR provides several different interrupt sources. These interrupts have separate vector locations in the program memory space. The lowest addresses in the program memory space are, by default, defined as the interrupt vectors. The lowest address location (0x0000) is allotted to the reset vector, which is not exactly an interrupt source. The address of an interrupt also determines its priority. The lower the address, the higher its priority level. So, reset has the highest priority. When two or more interrupts occur at the same time, the interrupt with the higher priority is executed first, followed by the interrupt with lower priority. Interrupts are used to suspend the normal execution of the main program and take the program counter to the subroutine known as the interrupt service routine (ISR). After the ISR is executed, the program counter returns to the main loop. The following illustration shows how the code in an ISR is executed.

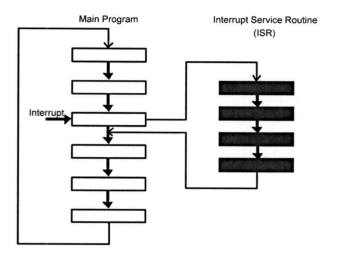

All interrupts are assigned individual enable bits, which must be set to logic one (as is the global interrupt enable bit in the status register) in order to enable the interrupt. When an ISR is executing, the global interrupt enable bit is cleared by default, and hence, no furthers interrupts are possible—unless the user program has specifically enabled the global interrupt enable bit to allow nested interrupts, that is, an interrupt within another interrupt. Various peripherals of AVR devices like timers, USI, ADC, analog comparator, etc., have different interrupt sources for different states of their values or status.

USI: Universal Serial Interface

The universal serial interface, or USI, provides the basic hardware resources needed for serial communication. This interface can be configured to follow either a three-wire protocol, which is compliant with the serial peripheral interface (SPI), or a two-wire protocol, which is compliant with the two-wire interface (TWI). Combined with a minimum of control software, the USI allows significantly higher transfer rates and uses less code space than solutions based on software only. Interrupts are included to minimize the processor load.

Analog Comparator

AVR devices provide a comparator, which measures the analog input voltage on two of its terminals and gives digital output logic (0 or 1), depending on whether the voltage on the positive terminal is high or that on the negative terminal is high. The positive and negative terminals can be selected from different I/O pins. The change in output of the comparator can be used as an interrupt source. The output of the comparator is available on the analog comparator output (ACO) pin. The following illustration shows the block diagram of the analog comparator.

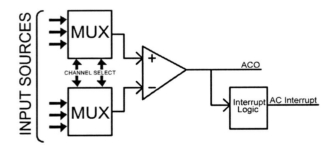

Analog to Digital Converter

These devices have a ten-bit, successive approximation–type ADC with multiple single-ended input channels. Some devices also have differential channels to convert analog voltage differences between two points into a digital value. In some devices, to increase the resolution of measurement, there is a provision to amplify the input voltage before conversion occurs. The reference voltage for measurement can be configured to be taken from the AREF pin, VCC, and the internal bandgap references. The following illustration shows the block diagram of the ADC.

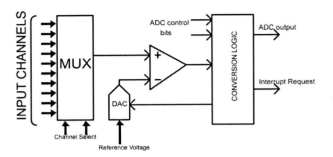

Clock Options

The system clock sources in the AVR devices include the calibrated resistor capacitor (RC) oscillator, the external clock, crystal oscillator, watchdog oscillator, low-frequency crystal oscillator, and phase lock loop (PLL) oscillator. The main clock can be selected to be any one of these through the fuse bits. The selected main clock can be further prescaled by setting suitable bits in the clock prescaler register during the initialization part of the user software. The selected main clock is distributed to various modules like CPU, I/O, Flash, and ADC.

- **CLK_CPU**   It is routed to parts of the system concerned with the operation of the AVR core, like register file, status register, etc.

- **CLK_I/O**   It is used by the majority of the I/O modules, like timer/counter, USI and synchronous external interrupts, etc.

- **CLK_FLASH**   The Flash clock controls operation of the Flash interface.

- **CLK_ADC**   Unlike other I/O modules, the ADC is provided with a dedicated clock so that other clocks can be halted to reduce the noise generated by digital circuitry while running the ADC. This gives more accurate ADC conversion results. The following illustration shows the various clock options.

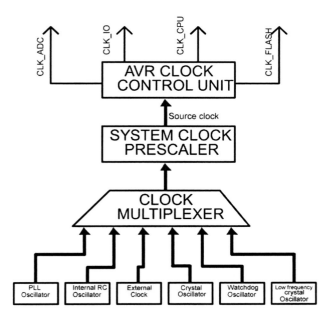

Power Management and Sleep Modes

It is necessary for the modern generation of controllers to manage their power resources in the utmost efficient manner, and AVR devices cannot afford to lag behind in this race of optimization. They support certain sleep modes, which can be configured by user software and allow the user to shut down unused modules, thereby saving power.

The sleep modes supported include power down, power save, idle, ADC noise reduction, etc. Different devices support different modes, and the details can always be found in the datasheets.

Furthermore, each mode has a different set of wakeup sources to come out of that mode and go to full running state.

System Reset

AVR devices can be reset by various sources, summarized here:

- **Power-on reset**   The microcontroller unit (MCU) is reset when the supply voltage is below the power-on reset threshold.

- **External reset**   The MCU is reset when a low level is present on the RESET pin.

- **Watchdog reset**   The MCU is reset when the watchdog is enabled and the watchdog timer period expires.

- **Brown-out reset**   The MCU is reset when the brown-out detector is enabled and the supply voltage VCC is below the brown-out reset threshold.

After reset, the source can be found by software by checking the individual bits of the MCU status register. During reset, all I/O registers are set to their initial values, and the program starts execution from the reset vector. The following illustration shows the block diagram of various reset sources.

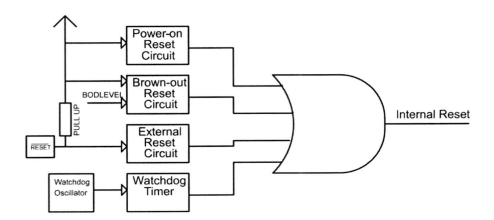

Memory Programming

Programming the AVR device involves setting the lock bits, setting the fuse bytes, programming the Flash, and programming the internal EEPROM. This data can also be read back from the controller along with signature bytes for identification of the device. Tiny devices can be programmed using serial programming or high-voltage parallel programming. Unless otherwise mentioned, throughout this book we have used serial programming for the Tiny microcontrollers. This method can be further divided into two other methods: in-system programming (ISP) and high-voltage serial programming (HVSP). HVSP is only applicable to eight-pin microcontrollers as an alternative to parallel programming, because these devices have too few pins to use parallel programming.

In-system programming uses the AVR internal serial peripheral interface (SPI) to download code into the Flash and EEPROM memory segments of the AVR. It also programs the lock bits and fuse bytes. ISP programming requires only VCC, GND, RESET, and three signal lines for programming. There are certain cases when the RESET pin must be used for I/O or other purposes. If the RESET pin is configured to be I/O (through the RSTDISBL fuse bit), ISP programming is unavailable and the device has to be programmed through parallel programming or high-voltage serial programming, whichever is applicable.

There is one more method to program these devices—the debugWIRE on-chip debug system, which is described in the next section. The recent series of six-pin devices from Atmel—ATtiny 4/5/9/10—doesn't support any of the previously mentioned methods of programming, but has a new tiny programming interface (TPI) built in for programming.

The lock bits are used for protection of the user software in order to prevent duplicity, and fuse bytes are used for initial settings of the controller that cannot and should not be performed by user software. The following illustration shows the signals for ISP serial programming.

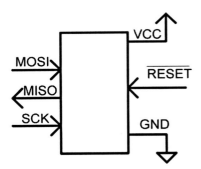

DebugWIRE On-Chip Debug System

The debugWIRE on-chip debug system is a one-wire interface for hardware debugging and programming the Flash and EEPROM memories. This interface is enabled by programming the debugWIRE enable (DWEN) fuse. After enabling this interface, the RESET pin becomes the communication gateway between the target and emulator. Thus, external reset doesn't work if this interface is enabled. This interface uses the same protocol as that used by JTAG ICE mkII, a popular debug tool from Atmel. The following illustration shows the debug WIRE interface.

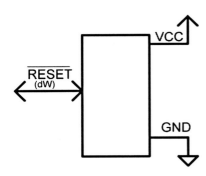

Elements of a Project

This book shows several projects spanning a wide spectrum of ideas and involving several application domains. These projects can be built for fun as well as education. However, it is important to dwell upon the design and development process.

How does one go about making a system or a project that no one has thought of before? Of course, you have to think what you need. Sometimes, the trigger for this need might come by looking at other people's projects. It's an abstract process, but an example might help to illustrate it. Suppose you saw LEDs being used in some system: bright, blinking LEDs that capture your imagination, and you think, hey! what if I could have these pretty LEDs on my cap in some pattern and make them blink or change intensity? This idea for something unique is the most important thing. The illustration on this page shows the design and development process.

Once an idea germinates in your mind, you can continue to evolve it. At the same time, an Internet search is recommended to ensure that no one else has already thought of the same idea. There is no point in reinventing the wheel. If the idea has been already implemented, maybe it would be good to think how it can be further improved. If you do indeed take up the implementation and improve upon it, a good plan of action would be to share it with the original source of the implementation, so as to acknowledge the work and also to put on record your own contribution. This way, one can enrich the system by contributing back to it. These ideas apply to projects that are available on the Internet under some sort of "freeware" license. In other cases, you may need to check up on the appropriate thing to do. It would be all right in most cases if you intend to use the original or your adaptation for personal use. If you intend to use it for commercial applications, however, it is absolutely necessary to check with the original source to avoid future problems.

There are two distinct elements in a project, as seen in the illustration, namely the hardware components and the software. The hardware part can be implemented in many ways, but using a microcontroller is an easy option, and since this book is about using microcontrollers in projects, that is what we are going to concentrate on. Apart from the microcontroller, the system needs a source of power to operate. It would also need additional hardware components specific to the project even though modern microcontrollers integrate a lot of features, as seen in the next illustration. For example, even though a microcontroller has digital output pins to control a bank of seven-segment displays, it does not have the capability to provide the large enough current that may be needed, so you will have to provide external current drivers. Similarly, if you want to

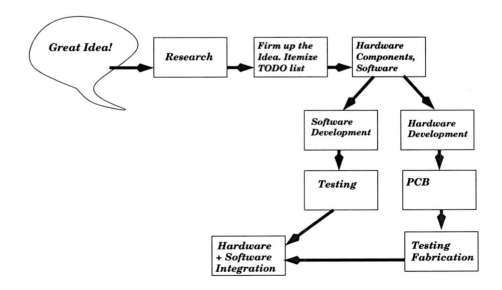

use an external sensor that provides an analog voltage to measure a physical parameter, the voltage range from the sensor may not be appropriate for use with the microcontroller's on-board ADC, so you would need an external amplifier to provide gain to the sensor output voltage. The illustration on this page shows the elements of a modern microcontroller.

The software component refers to the application program that runs on the microcontroller, but may also refer to a custom program that runs on a PC, for example, to communicate with the microcontroller.

The project development process requires that the two elements of the project, the hardware elements and the software elements, be developed in parallel. The software component that runs on the microcontroller is developed on a host PC, and a large section of the code can be developed even without the hardware prototype completed. The

software code can be tested on the PC host for logical errors, etc. Some parts of the code that require external signals or synchronization with other hardware events cannot be tested, and this testing must be postponed until the software is integrated with the hardware. Once the hardware prototype is ready, it must be integrated with the software part and the integrated version of the project tested for compliance with the requirements. The integration may not be smooth and may require several iterative development cycles.

Apart from the hardware components, which would be specific to a given project and the software, some hardware components are common across most projects. These are related to the power supply and a clock source for the microcontroller. These elements of the project are shown in the next illustration. The power supply source and the regulation of the supply voltage are discussed in detail in a later section. The clock

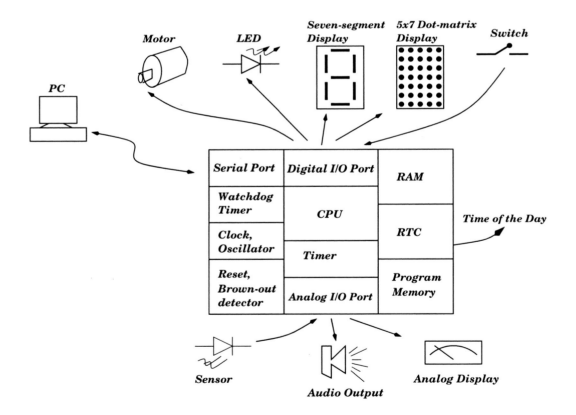

source is critical to the operation of the project. Fortunately, some sort of clock source is often integrated in the microcontroller itself. This is usually an RC oscillator that is not very accurate and whose actual value depends on the operating voltage, but is quite suitable for many applications. Only if the application requires critical time measurements does one need to hook up an external clock oscillator. All of the microcontrollers in the AVR family have an on-chip clock source, and in most projects in this book, we use the same. The rate of program execution is directly dependent upon the clock frequency; a high clock frequency means your program executes faster. However, a high clock frequency also has a downside: the system consumes more power. There is a linear dependence of power and clock frequency. If you double the clock frequency, the power consumption would also double. So, it is not very wise to choose the highest available frequency of operation, but rather to determine the frequency based on the program execution rate requirement. As we illustrate in Project 1 later in this chapter, by choosing to use the lowest available clock frequency, we are able to keep the required operating power to a minimal level. The following illustration shows the elements of a project.

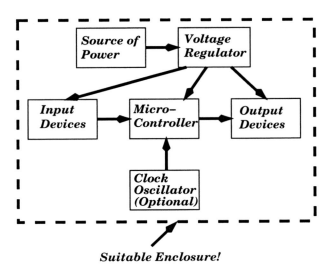

Suitable Enclosure!

Apart from the clock source, power supply source, and voltage regulator, the project requires input and output devices and a suitable enclosure for housing the project, as shown in the illustration.

Power Sources

For any system to run, a power supply is needed. Without the required supply, the system is only as good as a paperweight. Selecting the right source of power is important. For a portable system, connecting it to the main grid would tie it up to a physical location, and it would hardly be classified as a portable system then.

Batteries

Batteries are the most common source of energy for portable electronics applications. They are available in a variety of types, packages, and energy ratings. The energy rating of a battery refers to the amount of energy stored in it. Most batteries are of two types: primary and secondary. Primary batteries are disposable batteries. These batteries can provide energy as soon as they are assembled and continue to provide energy through their lifetimes or until they are discharged. They cannot be recharged and must be discarded. Secondary batteries, on the other hand, need to be charged before they can be used. They can be recharged several times in their usable lifetime and, therefore, are preferred over primary batteries, although secondary batteries are more expensive. Also, the energy density of a primary battery is better than that of a secondary battery. Energy density refers to the amount of energy stored in a battery per unit weight. So a primary battery with the same weight as a secondary battery can provide operating voltage for a longer time than the secondary battery can.

A popular primary battery is the zinc-carbon battery. In a zinc-carbon battery, the container is made out of zinc, which also serves as the negative terminal of the battery. The container is filled with a paste of zinc chloride and ammonium chloride, which serves as the electrolyte. The positive terminal of the battery is a carbon or graphite rod surrounded by a mixture of manganese dioxide and carbon powder. As the battery is used, the zinc container becomes thinner and thinner due to the chemical reaction (leading to the oxidation of zinc) and eventually the electrolyte starts to leak out of the zinc container. Zinc-carbon batteries are also the cheapest primary batteries. Another popular primary battery is the alkaline battery. Alkaline batteries are similar to zinc-carbon batteries, but the difference is that alkaline batteries use potassium hydroxide as an electrolyte rather than ammonium chloride or zinc chloride. Figure 1-3 shows some alkaline batteries. The nominal open circuit voltage of zinc-carbon and alkaline batteries is 1.5 volts.

Other common primary battery chemistries include the silver oxide and lithium variant. The silver oxide battery offers superior performance compared to the zinc chloride battery in terms of energy density. It has an open circuit terminal voltage of 1.8 volts. The lithium battery, on the other hand, uses a variety of chemical compounds, and depending upon these compounds, it has an open circuit terminal voltage between 1.5 and 3.7 volts. Figure 1-4 shows lithium and alkaline batteries in the form of button cells.

The only issue with primary batteries is that once the charge in the battery is consumed, it must be disposed of safely. This is where the use of secondary batteries looks very attractive: they can be recharged several times before you need to dispose of them. Rechargeable batteries are available in standard as well as custom sizes and shapes. Common rechargeable batteries are lead-acid, Ni-Cd, NiMH, and lithium-ion batteries. Figure 1-5 shows a lithium-ion battery. Charging these batteries requires a specialized charger, and only a suitable charger should be used with a particular battery. Charging a lithium-ion battery with a battery charger meant for, say, NiMH batteries, is not advisable and would certainly

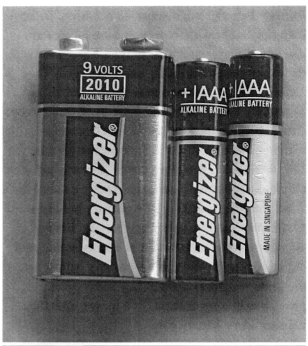

Figure 1-3 Alkaline battery in 9V- and AAA-size packages

Figure 1-4 The smaller LR44 cell is an alkaline battery. The bigger CR2032 cell is a lithium battery.

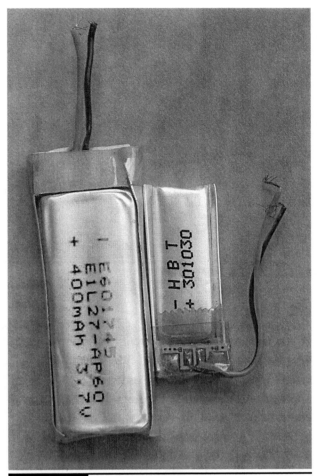

Figure 1-5 Lithium-ion battery

damage the battery as well as lead to the possibility of fire or battery explosion.

Primary and rechargeable batteries are available in many standard sizes. A few of the more common ones are listed in Table 1-2.

When selecting a battery for your application, the following issues need to be considered:

- **Energy content or capacity** This is expressed in Ah (or mAh) (ampere hour or milliampere hour). This is an important characteristic that indicates how long the battery can last before it discharges and becomes useless. For a given battery type, the capacity also dictates the battery size. A battery with a larger Ah rating will necessarily be bigger in volume than a similar battery with a smaller Ah rating.

- **Voltage** The voltage provided by the battery.

- **Storage** This indicates how the battery needs to be stored when not being used.

- **Shelf life** This indicates how long the battery will last before it discharges on its own. There is no point in buying a stock of batteries for the next ten years if the shelf life of the batteries is, say, only one year.

- **Operating temperature** Batteries have notoriously poor temperature characteristics. This is because the batteries depend upon a chemical reaction to produce power and the chemical reaction is temperature dependent. Batteries perform rather poorly at low temperatures.

- **Duty cycle** Some batteries perform for a longer period if they are used intermittently. The duty cycle of the battery indicates if the battery can be used continuously or not, without loss of performance.

TABLE 1-2	Battery Nomenclature and Dimensions				
Nomenclature	Shape	Length	Diameter/Width	Height
AAA	Cylinder	44.5 mm	12 mm	—
AA	Cylinder	50.5 mm	14.5 mm	—
9V	Rectangular cuboid	48.5 mm	17.5 mm	26.5 mm
C	Cylinder	50 mm	26.2 mm	—
D	Cylinder	61.5 mm	34.2 mm	—

Fruit Battery

Some of the fruits and vegetables we eat can be used to make electricity. The electrolytes in many fruits and vegetables, together with electrodes made of various metals, can be used to make primary cells. One of the most easily available fruits, the lemon, can be used to make a fruit cell together with copper and zinc electrodes. The terminal voltage produced by such a cell is about 0.9V. The amount of current produced by such a cell depends on the surface area of the electrodes in contact with the electrolyte as well as the quality/type of electrolyte.

Preparing the Battery

For the battery, we need a few lemons for the electrolyte and pieces of copper and zinc to form the electrodes. For the copper, we just use a bare printed circuit board (PCB), and for the zinc we chose to use zinc strips extracted from a 1.5V battery.

1. Start with a piece of bare PCB. The size of the PCB should be large enough so that you can create three or four islands on it. Each island will be used to hold a half-cut lemon.

2. Next, open up a few 1.5V AA size cells for the zinc strips and clean them up with sandpaper. Solder wire to each strip. Instead of these zinc strips, you can also use household nails. Nails are galvanized with zinc and can be easily used for making the battery.

3. On the bare copper PCB, cut islands with a file or hacksaw and solder the other end of the wire from the zinc strip to each copper island. For each cell, you need half a lemon, one island of copper, and one zinc strip.

4. Place the lemons on each copper island with the cut facedown as seen in Figure 1-6. Make incisions in the lemons to insert the zinc strips. The photograph in Figure 1-6 shows a lemon battery with four cells.

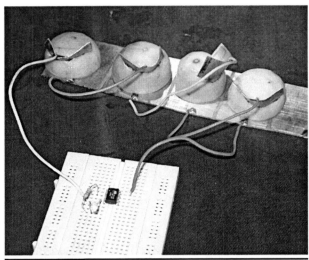

Figure 1-6   Lemon battery

AC Adapter

If you use an alternating current (AC) output adapter, then the rectifier and filter capacitor circuit must be built into the embedded application, as shown in Figure 1-7. The rectifier could be built with discrete rectifier diodes (such as 1N4001), or a complete rectifier unit could be used. The rectifier should be suitably rated, keeping in mind the current requirements. If the power supply unit is to provide 500mA of current, the diodes should be rated at least 1A. The other rating of the diode to consider is the PIV (peak inverse voltage). This is the maximum peak reverse voltage that the diode can withstand before breaking down. A 1N4001 diode has a PIV of 50V, and 1N4007 is rated to 1000V.

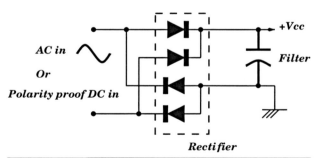

Figure 1-7   Rectifier and filter capacitor circuit: It can be used with AC input as well as DC input voltage.

The peak rectified voltage that appears at the filter capacitor is 1.4 times the AC input voltage (AC input voltage is a root mean square [RMS] figure). A 10V AC will generate about 14V direct current (DC) voltage on the filter capacitor. The filter capacitor must be of sufficiently large capacity to provide sustained current. The filter capacitor must also be rated to handle the DC voltage. For a 14V DC, at least a 25V rating capacitor should be employed. The rectifier filter circuit shown in Figure 1-7 can also be used with a DC input voltage. With this arrangement, it would not matter what polarity of the DC voltage is applied to the input of the circuit.

Once raw DC voltage is available, it must be regulated before powering the embedded application. Integrated voltage regulator circuits are available. Voltage regulators are broadly classified as linear or switching. The switching regulators are of two types: step up or step down. We shall look at some of the voltage regulators, especially the so-called micropower regulators.

It is common to use the 78XX type of three-terminal regulator. This regulator is made by scores of companies and is available in many package options. To power the AVR processor, you would choose the 7805 regulator for 5V output voltage. It can provide up to 1A output current and can be fed a DC input voltage between 9V and 20V. You could also choose an LM317 three-terminal variable voltage regulator and adjust the output voltage to 1.25V and above with the help of two resistors.

A voltage regulator is an active component, and when you use this to provide a stable output voltage, it also consumes some current. This current is on the order of tens of milliamperes and is called the quiescent or bias current. Micropower regulators are special voltage regulators that have extremely low quiescent current. The LP2950 and LP2951 are linear, micropower voltage regulators from National Semiconductor, with very low quiescent current (75mA typ.) and very low dropout voltage (typ. 40mV at light loads and 380mV at 100mA maximum current). They are ideally suited for use in battery-powered applications. Furthermore, the quiescent current of the LP2950/LP2951 increases only slightly at higher dropout voltages. These are the most popular three-terminal micropower regulators, and we use them in many of the projects.

USB

The Universal Serial Bus (USB) is a popular and now ubiquitous interface. It is available on PCs and laptop computers. It is primarily used for communication between the PC as the host and peripheral devices such as a camera, keyboard, etc. The USB is a four-wire interface with two wires for power supply and the other two for data communication. The power supply on the USB is provided by the host PC (or laptop or netbook). The nominal voltage is +5V, but is in the range of +4.4V to +5.25V for the USB 2.0 specifications. The purpose of providing a power supply on the USB is to provide power to the external devices that wish to connect to and communicate with the PC. For example, a mouse requires a power supply for operation and it can use the USB power. However, this voltage can be used to power external devices also, even if the device is not going to be used by the PC. We use USB power to provide operating voltage to an embedded application, especially if it is going to be operated in the vicinity of a PC or laptop. The embedded circuit can draw up to 100mA from the USB connector; although the USB can provide larger current, it cannot do so without negotiation (i.e., a request) by the device. Table 1-3 shows the pins of the USB port that provide power and signal.

TABLE 1-3	Pins of the USB Mini- or Microconnector		
Pin	Name	Connecting Wire Color	Purpose
1	Vcc	Red	+5V
2	D	White	Data signal –ve
3	D+	Green	Data signal +ve
4	ID	None	Device identification
5	Gnd	Black	Ground

Solar Power

Solar energy could be used to power electronic circuits by using photovoltaic cells. They provide power as long as the cell is exposed to sunlight. Solar cells provide a range of power, from less than a watt to hundreds of watts. The output power of a solar cell is directly proportional to the incident light and inversely proportional to the cell temperature. To ensure maximum ambient light, the solar cell must be held perpendicular to the incident light. A conversion circuit is often used to regulate the output of the cell. The most common use of a solar cell is to charge a battery so that continuous power from the battery can be derived. More details on the use of solar cells are covered in a later chapter.

Faraday-based Generator

The operating voltage required for many small embedded portable projects can be met by an interesting device that converts mechanical energy into electrical energy. This uses the famous Faraday's law. The device based on this principle is shown in Figure 1-8. The system uses a hollow Perspex tube of suitable diameter and length. Inside the tube is placed a rare earth magnet. The tube is wound with several hundred turns of copper enameled wire. The ends of the tube are sealed. To generate the voltage, the tube is simply shaken. As the magnet traverses the length of the tube, it

produces AC voltage across the copper wire, which can be rectified and filtered using the circuit shown in Figure 1-7 to provide DC voltage. The only issue with this method is you have to keep shaking the tube for as long as you want to power the circuit. Once you stop shaking the tube, it will stop producing the voltage and only the residual voltage on the capacitor will be available. In many applications, this may not be an issue. One possible solution is to use supercapacitors instead of normal capacitors. However, it would take a long time and a lot of effort to charge the supercapacitors to the required voltage.

The DC voltage produced at the capacitor terminals may further require a voltage regulator before the voltage is connected to the application circuit, and a low dropout and low quiescent voltage regulator such as the LP2950 is recommended.

The photograph in Figure 1-9 shows the output of the Faraday generator captured on an oscilloscope. The output is more than 17V peak to peak.

RF Scavenging

Radio frequency (RF) waves are ubiquitous, and therefore it is possible to receive the radio frequency energy using a suitable antenna and convert this to DC operating voltage. Unfortunately, this scheme requires a large transmitted power from the source, or a large antenna, or close proximity to the source. In many

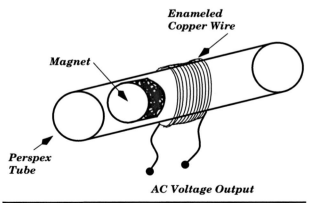

Figure 1-8   Faraday-based voltage generator

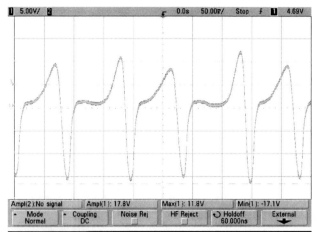

| Ampl(2):No signal | Ampl(1): 17.8V | Max(1): 11.8V | Min(1): -17.1V |
| Mode Normal | Coupling DC | Noise Rej | HF Reject | Holdoff 60.000ns | External |

Figure 1-9   Output of a Faraday generator

commercial applications, the RF energy is deliberately transmitted for use by such receivers. One such application is the radio frequency identification device (RFID) systems. The block diagram of such a system is shown in Figure 1-10.

The system consists of an unmodulated radio frequency transmitter transmitting RF power at a suitable frequency. The frequency of operation is determined by the quartz crystal used. A higher frequency of operation would require a smaller transmission antenna. The transmitter is powered with a DC supply voltage of a suitable value. The radiated signal is received by a tuned circuit consisting of an inductor and a variable capacitor in parallel that is tuned to the frequency of the transmitter. The tuned circuit feeds a diode rectifier, filter, and a suitable low-power voltage regulator. The output of the regulator provides the operating supply voltage to the desired circuit. Such a system can provide few milliwatts of power across distances in the range of few tens of centimeters.

A practical system based on this approach is described in the following EDN Design Idea: "Wireless battery energizes low-power devices": www.edn.com/article/CA6501085.html.

Hardware Development Tools

To develop and make prototypes for the projects described in this book, we have used some commonly available tools. These tools are:

- **Solder iron, 35 watts, with a fine solder tip**   A soldering station is highly recommended, but is not mandatory. The soldering station offers isolated supply to the solder iron heater, thus reducing the leakage currents from the tip of the solder iron.

- **Solder wire**   A thin-gauge solder wire is recommended. We use 26 SWG solder wire. The photograph in Figure 1-11 shows the solder wire and iron.

- **Copper braid**   This is often useful in desoldering components.

- **Eye loupe**   To inspect PCBs, solder joints, etc. Eye loupe and copper braid are shown in Figure 1-12.

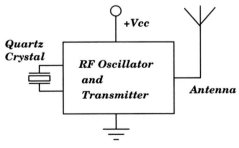

Power Broadcasting Circuit

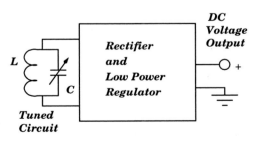

Power Receiver Circuit

Figure 1-10   Power supply from a radio frequency source

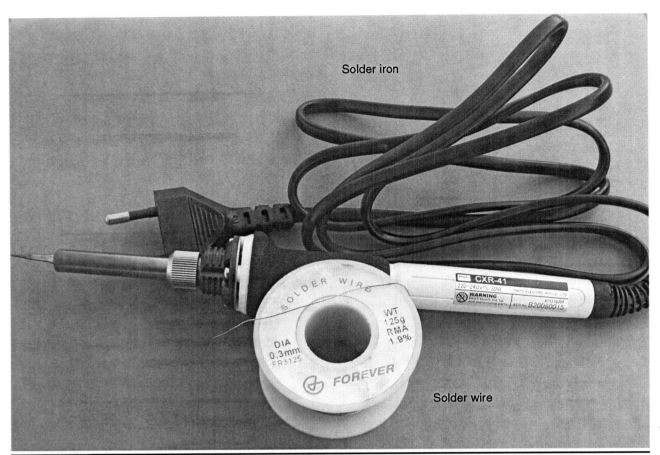

Figure 1-11 Solder wire and solder iron

Figure 1-12 Copper braid and eye loupe

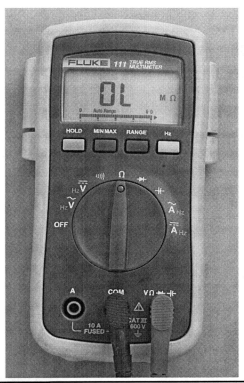

Figure 1-13 Multimeter

- **Multimeter** A digital multimeter with voltage, current, and resistance measurement facilities is useful for testing and measurements. It is shown in Figure 1-13.

- **Fine tweezers** For bending component leads.

- **Nipper** To cut the component leads. This is a fancy name for the regular lead cutter. A nipper has sharp edges that make a neat cut.

- **Needle-nose pliers** Generally useful for tightening screws, etc.

- **Screwdriver set** Tweezers, nipper, needle-nose pliers, and screwdriver set are shown in Figure 1-14.

- **M3 nuts and bolts** For fastening brackets onto the PCB as well as to support the PCB.

- **Drill machine (hand operated will do), with an assorted collection of drill bits** Used for drilling holes in the PCB, enclosures, etc.

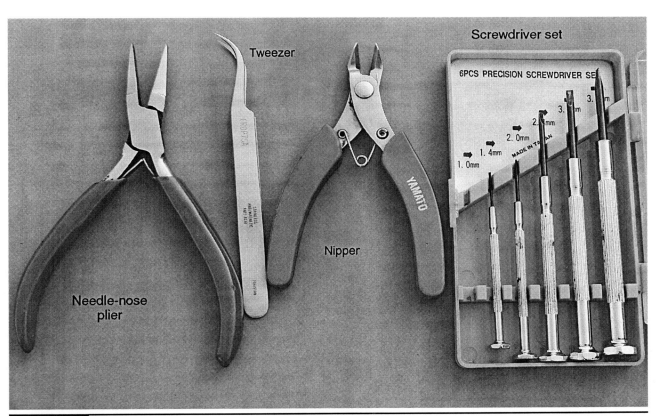

Figure 1-14 More tools

Figure 1-15 Bench vice

- **Bench vice with a three-inch jaw** For holding the PCB steady, filing hardware or PCB, etc. It is shown in Figure 1-15.

Software Development

The advantage of developing a programmable system cannot be realized without writing efficient code for the programmable devices of your system, in this case, tinyAVR microcontrollers. Throughout this book, we will use C language for programming them. The syntax is in compliance with GNU's AVR-GCC compiler.

C is a high-level programming language, and code written in C language has to be converted into machine language that your target controller understands and can execute. The tool that does this conversion is called a compiler. The Tiny controllers understand only binary format and have to be fed in bytes. A common way of storing a collection of bytes to be transferred to the controller is to use a hex file that contains the bytes in the form of hexadecimal notation. So there must be a tool that can convert C code into the hex file. Many varieties of C compilers for AVR microcontrollers are available, but we have focused on the AVR-GCC for obvious reasons. WinAVR

gives a good integrated development environment (IDE) for using AVR-GCC on Windows.

WinAVR, apart from giving you nice tutorials on the AVR C library, provides the following two main utilities:

- **Programmer's Notepad** It's a general-purpose IDE for programming in numerous languages. This software comes integrated with the WinAVR compiler. To run Programmer's Notepad, go to Windows | Programs | WinAVR (version) | Programmers Notepad. Figure 1-16 is the screen shot of Programmer's Notepad. As you can see, it has various tabs. The most important tab, Tools, is shown displayed. As you can see, it has three important commands:

  - **Make All** To compile the program by running the MAKEFILE and generate the hex file.

  - **Make Clean** To remove all the hex files and other dependencies. Generally used before recompiling the program.

  - **Make Program** This can be used for burning your hex file into the microcontroller, but it requires a special-purpose ISP programmer.

- **MAKEFILE Template** Converting your C code into the hex files involves numerous tasks like preprocessing, compiling, linking, and finally loading. GCC (GNU C compiler) compilers generally require commands to be given for each process to be carried out. If you give all the commands each time, one by one, when you compile your code, your task would become cumbersome. In this situation, a utility called MAKEFILE helps. It integrates all its commands in one place and does the complete job by giving instructions one by one to the compiler. WinAVR gives you the basic MAKEFILE template, which you can modify for your own needs. To run this, go to Windows | Programs | WinAVR (version) | mFile. Set your options and save the file. Note

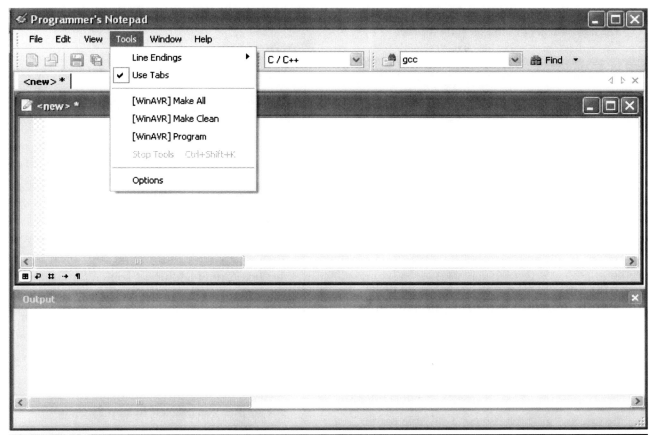

Figure 1-16 Programmer's Notepad

that making the MAKEFILE from scratch can be tough for beginners. Hence, if you are not comfortable with MAKEFILE options, you can use the MAKEFILE provided in the codes of this book with slight modifications to suit your needs.

Working with WinAVR and its components can be a little tricky during the initial stages. On the other hand, AVR Studio from Atmel allows easy management of C projects with automatic handling of make commands (required to compile the code written for the GCC compiler). However, AVR Studio still uses WinAVR GCC at the back end to compile your C language code, as it doesn't have a built-in C compiler and only offers a built-in assembler to run Assembly language programs. So you need to install both WinAVR GCC and AVR Studio to get started with programming. The latest

version of AVR Studio can be downloaded from http://www.atmel.com/dyn/Products/tools_card .asp?tool_id=2725 and that of WinAVR from http://sourceforge.net/projects/winavr/files. The projects in this book have been directly compiled through WinAVR's Programmer's Notepad, with manual handling of make commands through MAKEFILE. However, you can use either of the two methods. A quick-start introduction to Embedded C programming for AVR microcontrollers is given in Appendix A. The instructions for getting started with both methods are given next.

Getting Started with a Project on AVR Studio

To run AVR Studio, go to Windows I Programs I Atmel AVR Tools I AVR Studio 4.

1. To create a new project, select New Project from the Project menu as shown here:

2. After clicking New Project, a pop-up menu appears, as shown in the next illustration. In the Project Type field, select either AVR GCC or Atmel AVR Assembler, depending on the language to be used. Here, settings are shown for making a C language project. Select both the Create Initial File and Create Folder check boxes, and give the project a suitable name. Click Next.

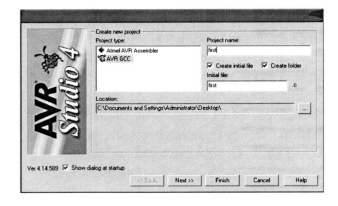

3. After clicking Next, a pop-up menu, shown next, appears. Click AVR Simulator, and from the Device section select the suitable controller. Click Finish, and you will see the main source file open and you can start writing your code.

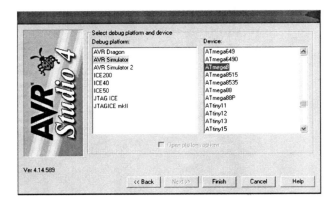

4. Often, you need to break your code into sections for portability and readability. So you divide your code into several code files. To include further additional source files, right-click Source Files in the AVR GCC section, and select either Add Existing Source File or Create New Source File, depending upon your requirement. If you are using existing source files, make sure that they are copied in the same folder as your main source file (the folder that you created in step 2).

5. Write your code in the main source file.

6. From the Build menu, select the Build command (or press F7) to start compilation of your program. If you see "Build succeeded with 0 Warnings" in the Build window, it means there is no error and your hex file has been created. If you see "Build succeeded" along with some warnings, it means that your hex file has been created but with some warnings. It is recommended that the source of the warnings be investigated to remove them, if possible. The hex file is located inside the subdirectory "default" in the main project folder.

7. You can also select the Build And Run command from the Build menu to simulate your program (or press CTRL-F7). The single instruction step-over is performed using the F11 key. During simulation, the contents of the controllers' register, I/O ports, and memory can also be monitored after each instruction execution.

Getting Started with a Project on WinAVR

To start a new project using WinAVR, the following steps should be followed:

1. Make a new folder at any location in your PC.

2. In this folder, copy the MAKEFILE from any project of this book (let's say Chapter 1). Experienced users can make their own MAKEFILE. This MAKEFILE will match most of your requirements. The critical locations where you may want to make a change in the MAKEFILE for different projects are shown here. Lines beginning with # are treated as comments in the MAKEFILE.

 # MCU name

MCU = your device

Example

 # MCU name

MCU = attiny861 (This tells the compiler that the microcontroller for which the application has to be compiled is Attiny861.)

#Output format. (Can be srec, ihex, binary)

FORMAT = ihex (The final output file has to be in hex format.)

 # Target file name (Without extension)

TARGET = main (This is the name of your hex file.)

 # List C source files here. (C dependencies are automatically generated.)

SRC = $(TARGET).c (This line shows the name of the source file. The command $(TARGET) is replaced by the value of TARGET that is main. Hence, you have to keep the name of your source file as main.c.)

 # If there is more than one source file,

 # append them above, or modify and

 # uncomment the following:

#SRC += abc.c

SRC += def.c

As explained earlier, you often have to break your code into several code files. To include additional source files, add them as shown here. In the previous example, abc.c is not included, as the line SRC += abc.c is commented out and def.c is included. You can create your own source files and add them here.

3. Next create an empty text document and name it main.c, as explained earlier.

4. Modify the MAKEFILE as per your needs.

5. Write your code in the main.c file.

6. From the Tools tab, click Make All. If you see the process exit code as 0, that means there is no error and your hex file has been created. If you see any other exit code, it means that there is an error and you must remove it. If the exit code is 0 and you see some warnings, the hex file is still created. As stated earlier, try to remove the warnings if possible. Sometimes, warnings during the code compilation lead to the project working erratically.

ANSI C vs. Embedded C

ANSI C is the standard published by the American National Standards Institute (ANSI) for the C programming language. Software developers generally follow this standard for writing portable codes that run on various operating systems. Even the original creator of C, Dennis Ritchie, has conformed to this standard in the second edition of his famous book, *C Programming Language* (Prentice Hall, 1988). When software developers write a C program for a personal computer, it is run on an operating system. When the program has finished, the operating system takes back control of the CPU and runs other programs on it that are in the queue. In case of multiprocessing (or multithreading) operating systems, many different programs are run simultaneously on a personal computer. This is achieved by time slicing, that is, allowing each program in the queue access to the CPU, memory, and I/O, one by one, for fixed or variable durations.

When a program completes, it is removed from the queue. This gives the impression that these programs are running simultaneously, but in reality, the CPU is executing only one sequence of instructions (a program) at a given time. This "scheduling" is controlled by the operating system, which keeps the main CPU occupied all the time.

In contrast to the previous approach, when one writes C code for low-end microcontrollers (although the AVR has advanced processor architecture, we are comparing it with the standard PC processors here), they are the only programs running on the hardware and have complete control of all resources. The usage of the operating system is not common in embedded systems. These programs generally run in infinite loops and do not terminate.

Hence, it is evident that the approaches to programming have to differ (to a certain extent) in both the cases. In spite of this fact, certain basic features of C programming, like data types, loops, control statements, functions, arrays, etc., are similar in ANSI C compilers and embedded C compilers.

Making Your Own PCB

Deciding on the software and hardware requirements for your project is all very well, but to actually implement your project, you need to get the circuits ready, with all of the components in their proper places for programming and testing. Beyond a certain number of components, prototyping on a breadboard no longer remains feasible. Circuits made on a breadboard have more chances of behaving in an erratic manner due to undesirable shorts, loose connections, or no connections. These problems can be difficult to debug, and you spend extra time solving these

problems rather than testing your hardware design idea and compatible software. The other stream of circuit designing involves augmenting your projects with the printed circuit boards (PCBs). Circuits developed on PCBs are more durable and less prone to failures compared to circuits made on breadboards. Once fabricated and soldered properly, you can be sure of all the connections and focus on more important areas—system design and software development. The process of PCB designing and fabrication can be divided into two broad categories, described in the following sections.

Using a General-Purpose Circuit Board

This approach is generally used by hobbyists and university students to quickly solder their circuits. Designing and fabricating a custom PCB takes up considerable amounts of time and resources, which are not available to everyone. A viable idea in such cases is to place your components on a general-purpose circuit board and then make the connections by soldering insulated copper wires. A single-strand, tinned copper wire, with Teflon insulation, is quite sturdy and flexible, and is a recommended option. The general-purpose circuit board has holes arranged on a 0.1-inch pitch with solder contacts on one side. These boards are of several types, and two popular types are synthetic resin bonded boards (called FR2) and glass epoxy boards (FR4). The former are cheaper but less durable than the latter. Hence, it is always a better idea to use the glass epoxy board. Figure 1-17 shows a bare general-purpose glass epoxy circuit board, and Figures 1-18 and 1-19 show two sides of a general-purpose board with a circuit soldered on it. Sometimes, it is a good to idea to test and verify small modules of your designs on this board before proceeding towards the custom PCB design of the whole circuit. Wiring errors on such a board can easily be corrected, as the connections are

Figure 1-17 General-purpose circuit board

Figure 1-18 General-purpose board— component side view

Figure 1-19 General-purpose board—solder side view

made using wires that can always be changed from one point to another. But in a custom PCB, once the tracks are laid down, it is very difficult to change the circuit connections.

Creating a Custom PCB

This is the more advanced method for designing circuits. Circuits made on PCBs have a minimum chance of failure due to faulty connections unless there is some problem in the design. The first step in making a custom PCB is to use custom software on a PC for schematic design, layout, and routing of the board. Many types of PCB designing

software are available, such as Proteus, Orcad, and EAGLE. Throughout this book, we have used the freeware version of EAGLE from CadSoft for designing the boards of all the projects. The boards have been routed so that the maximum tracks come in a single layer to suit the single-side fabrication, which is a lot cheaper and easier than double-sided PCB fabrication. A quick-start tutorial on designing PCBs using EAGLE is given in Appendix B.

The designed PCB can be fabricated by many methods. These fall into two broad categories— etching and milling processes. Etching methods generally use chemicals, photographic films, silkscreens, etc., to remove the unwanted copper from the bare PCB while retaining the necessary tracks, whereas milling processes physically mill away the extra copper while leaving the copper for tracks intact. Milling operations, unlike etching processes, can be performed directly with the PCB designing software. Most of the PCBs in this book have been fabricated using a Roland Modela MDX-20 PCB milling machine. This machine supports single-side PCB fabrication, and all of the PCBs for our projects here have been adapted to suit the requirements of this machine. Milling processes are slow for mass production, because every PCB fabrication cycle has to be repeated from the start, but very quick for prototyping. Etching processes tend to be faster for mass production due to the reusability of intermediate films but are expensive and slow when the number of units required is small.

As a hobbyist, you may not be interested in getting the custom PCBs made. But there are often cases when you want multiple copies of your circuit, or wiring a circuit on a general-purpose board is cumbersome due to the complexity and large number of connections, or you have to submit your project for further evaluation. In such cases, custom PCBs are the only alternative.

Project 1
Hello World! of Microcontrollers

Now that we have described all the elements and components of a project, along with specific requirements to design the projects with AVR microcontrollers, here is a simple project for illustration. It has all the elements as shown in the illustration at the end of the "Elements of a Project" section earlier in the chapter. There are two LEDs and two switches in the circuit along with a reset switch. The aim of the project is to toggle the state of each LED each time the corresponding switch is pressed and released. The project is named as such because it introduces you to the world of tinyAVR microcontrollers.

Figures 1-20 and 1-21 show the schematic diagram used for this introductory project. Both schematics are identical, and they illustrate two popular styles of drawing a schematic. In the method illustrated in Figure 1-20, all connections between various component signals are explicitly shown using connecting lines. In the alternate style of drawing a schematic as illustrated by Figure 1-21, the signals are assigned signal names, such as PB3, which happens to be pin 2 of the

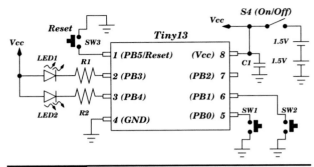

Figure 1-20 Hello World!

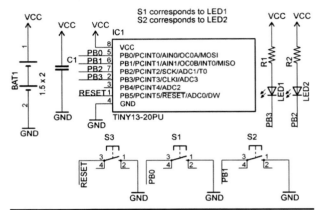

Figure 1-21 Hello World! alternate schematic style

microcontroller. Pin 2 is supposed to connect to LED1. So, signal name PB3 is assigned to pin 2 as well as to the cathode of LED1. Similar signal names are used for the rest of the connections.

Let us correlate the elements in the illustration with the components in the schematic shown in Figure 1-20 (or Figure 1-21). The circuit is powered with two AA alkaline batteries. As mentioned in the previous section, alkaline batteries have a nominal terminal voltage of 1.5V. Thus, the two batteries provide 3V operating voltage. Tiny13V operating voltage range is between 1.8V and 5.5V, so a 3V operating supply voltage would work fine. Also, as the batteries discharge, the terminal voltage would drop but the circuit would continue to work until the supply voltage drops to 1.8V. Also, the visible-spectrum LEDs (as opposed to invisible LEDs such as infrared) have a turn-on voltage, depending on the color of the LED between 1.8V (red) and 3.5V (white). Thus, selecting red LEDs for this project would be a good decision. The board layouts are shown in Figures 1-22 and 1-23. Figure 1-22 shows the layout without the tracks in the component layer (top), and Figure 1-23 shows the layout without the tracks in the solder (bottom) layer. As you can see, the board is mainly routed in the solder layer, with just one jumper in the component layer. It can easily be made using the physical milling process described in the previous section. The soldered prototype is shown in Figure 1-24.

The board layout in Eagle, along with the schematic, can be downloaded from www.avrgenius.com/tinyavr1.

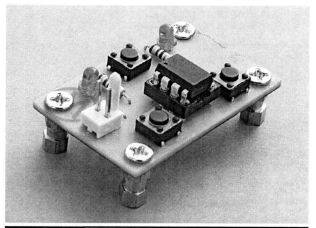

Figure 1-24 Hello World! soldered board

The code has been written in a way that the left switch toggles the left LED and the right switch toggles the right LED. Thus, if the right LED is off and you press and release the right switch momentarily, it will turn the right LED on. The Tiny13V is programmed with the following C code:

```c
//Include Files
#include<avr/io.h>
#define F_CPU 128000UL
#include<util/delay.h>

int main(void)
{
  DDRB  |= 1<<2|1<<3;//Declare as outputs
  PORTB |= 1<<2|1<<3;
    //Switch off the LEDs
  DDRB &= ~(1<<0|1<<1);//Input declared
  PORTB |= (1<<0|1<<1);//Pull up Enabled
  while(1)
  {
  //switch1
  if(!(PINB&(1<<0))) //If pressed
  {
    _delay_ms(10);//debounce
    while(!(PINB&(1<<0)));
      //wait for release
    _delay_ms(10);//debounce
    PORTB^= (1<<3);//Toggle
  }
  //switch2
  if(!(PINB&(1<<1)))//If pressed
  {
```

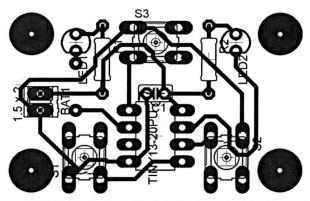

Figure 1-22 Hello World! PCB layout with solder layer shown

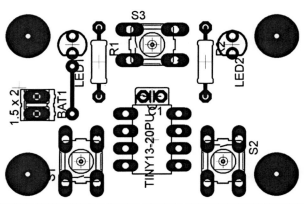

Figure 1-23 Hello World! PCB layout with component layer shown

(continued on next page)

```
    _delay_ms(10);//debounce
    while(!(PINB&(1<<1)));
        //wait for release
    _delay_ms(10);//debounce
    PORTB^= (1<<2);//Toggle
  }
 }
}
```

This code represents the general style followed in this book. The header files are specific to the AVR-GCC compiler. The macro F_CPU is used to convey the frequency of operation to the compiler. The program runs in an infinite loop. There is one single if block for each switch that first checks whether the switch is pressed or not. If the switch is pressed, it waits for it to be released, and on release, it performs the necessary action (toggling the LED). The 10ms delay after each switch press and release is for preventing switch bounce. Beginners who are new to C programming for AVR microcontrollers are advised to read Appendix A to better understand of this code. The compiled source code, along with the MAKEFILE, can be downloaded from www.avrgenius.com/tinyavr1.

The AVR is programmed using STK500 in ISP mode. The fuse settings are shown in Figure 1-25. The Tiny13 is set up to operate at 128 KHz internal RC clock. The idea is to clock the AVR at the lowest possible clock speed, since the power consumption of a complementary metal-oxide semiconductor (CMOS) digital circuit (such as this AVR) is directly proportional to the clock frequency of operation and we want to minimize the power dissipation.

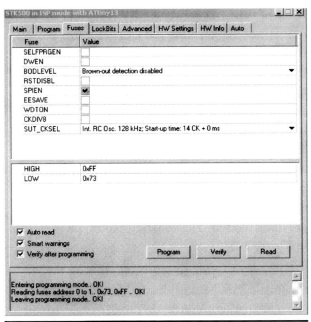

Figure 1-25 Fuse bits for the Tiny13 microcontroller

Conclusion

We have now reached the end of the first phase of this book and have covered a wide array of topics required for sound project development. This chapter covers all the basics for the following chapters. This chapter also has one beginner's project, and we have given the full source code and layout files here. The joyous ride of building creative projects with tinyAVRs begins with Chapter 2. For the rest of the projects, we haven't included the full source code and board layout in text. These are available on the links provided with each project. Crucial sections of the code, however, are explained. The next chapter shows a few simple LED-based projects.

LED Projects

IN THIS CHAPTER WE WILL describe a few simple projects that use LEDs. LEDs are popular electronic components, and recently, advances in technology have created LEDs that emit light spanning almost the entire visible spectrum. LEDs were invented in the early 1960s, with the earliest LEDs being red, which were followed by yellow ones. Although blue LEDs were demonstrated in the early 1970s, they weren't bright enough for practical use. Bright blue LEDs started becoming available by the late 1990s. The current focus is on white LEDs to provide lighting solutions. High-power and high-brightness white LEDs with 10W consumption providing 1,000 lumens of light are currently available. The main advantage of white LEDs for lighting is the promise of long life— 100,000 hours—as well as high efficiency. Thus, LEDs offer great opportunities for interesting projects, and in this chapter, we explore a few simple projects involving LEDs, including multicolor LEDs.

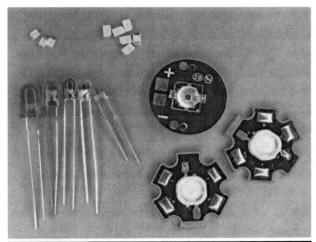

Figure 2-1 Various types of LEDs

LEDs

A light-emitting diode (LED) is a fascinating component. LEDs are available in a wide variety of sizes and colors. Figure 2-1 shows some of them. Many of the LEDs are available in clear packaging and so one cannot tell the color of the LED light simply by looking at the LED. An LED emits light when it is forward-biased and some current passes through it. The illustration here

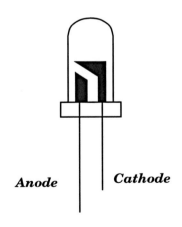

Anode　　　*Cathode*

Anode　　　*Cathode*

shows what a typical small-size LED looks like and also gives its electrical symbol. The two leads of the LED are not of equal size. The longer lead is the anode and the shorter one the cathode.

The intensity of the light emitted is proportional to the current passing through the LED. The forward voltage drop across the LED is dependent upon the semiconductor material that is used to make the LED, which is equivalent to saying that the forward voltage drop across the LED depends upon the color of the light produced by it. An LED that produces red light has the smallest bandgap (among the visible-light LEDs) compared to the blue light LEDs, which have the greatest bandgap. The forward voltage drop across a red LED is 2V and across a blue LED is 3.7V. Table 2-1 lists the LED color and typical electrical and optical characteristics.

Since an LED is a special diode, it is natural to expect the voltage and current characteristics to be identical to that of a normal signal diode. Signal diodes are made with silicon or germanium, and the turn-on voltage is around 0.7V (or 0.2V for germanium). The material used for making LEDs is gallium arsenide (GaAs), with suitable impurities added to get the desired light color. The band gap of GaAs is 1.45eV while that of silicon is 1.1eV. Therefore, the turn-on voltage of a diode made with GaAs is expected to be larger than the turn-on voltage of a silicon diode. The illustration here

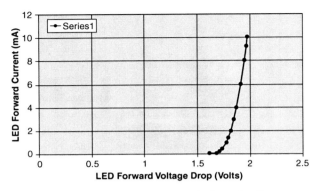

shows the forward-bias characteristics of a 5-mm red LED.

To use an LED, you need a suitable circuit and a DC voltage source to forward-bias the LED. The circuit is needed to determine the amount of current passing through the LED. For a particular LED, the forward current is chosen depending upon the intensity required from the LED and the amount of current that the LED can handle. One cannot exceed the current rating over extended periods without running a risk of damaging the LED. The peak forward current rating should never be exceeded. In its simplest form, the circuit associated with the LED to determine the current is just a simple resistor. For a supply voltage Vcc, the LED voltage drop of, say, V(f) across the LED and a desired current of 10mA, the value of the series resistor would be $R = \{Vcc - V(f)\}/10mA$.

As an example, if the LED in question is the one whose characteristics are plotted in the previous illustration and assuming Vcc = 5V and

TABLE 2-1	5-mm LED Electrical and Optical Characteristics (Lite-On Optoelectronics)				
Color	Forward Current I(av)	Peak Current I(pk)	Typical Forward Voltage V(led)	Viewing Angle	Wavelength
Red	20mA	120mA	2.0V	30	635nm
Orange	20mA	60mA	2.05V	15	624nm
Yellow	20mA	90mA	2.0V	20	591nm
Green	20mA	100mA	3.5V	15	504nm
Blue	20mA	100mA	3.7V	20	470nm
White	20mA	100mA	3.5V	20	Wide spectrum

V(f) from the curve is 1.98V, the value of the series resistor comes out to 302 Ohms. The standard 5% resistor series has 270 Ohms and 33 Ohms resistance, which can be put in series to get 303 Ohms, and that would work just fine. Instead of the series combination, if a single resistor of 270 Ohms is used, the LED current would increase to about 11mA, and if a 330 Ohm standard resistance is used, the current would be 9.15mA.

The intensity of light from an LED is proportional to the current passing through the LED when it is forward-biased. To illustrate this feature, the schematic in the image shown next was assembled and the readings were plotted. The result of this small experiment is shown in the plot in the second image. The significance of the relationship between intensity and current will be explored later in this section and in later projects.

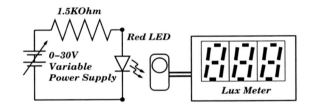

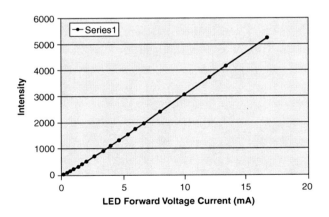

Types of LEDs

Apart from the LEDs shown in Figure 2-1, LEDs are also available in other forms, namely bicolor and red, green, blue (RGB) LEDs. The next

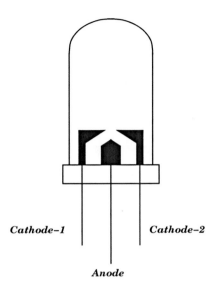

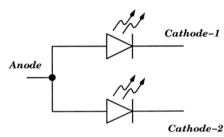

illustration shows a common anode bicolor LED and its electrical symbol.

Figure 2-2 shows a photograph of some bicolor and RGB LEDs. Bicolor LEDs in a common cathode configuration are also available. The two LEDs inside the package can be chosen from a wide variety of color combinations, such as red-yellow, red-green, blue-yellow, blue-red, etc. A single LED package with three color LEDs is also available. The colors are red, green, and blue, which are primary colors. In an RGB LED, there are four pins: a common anode and three cathodes, or a common cathode and three anodes. With a bicolor LED, multiple colors can also be formed by turning both LEDs on at the same time. So, with a red-green bicolor LED, you can get yellow by turning on the red and the green LEDs together. Similarly, if the red LED is turned on at half the intensity and the green is turned on at full intensity, then other shades of yellow can be

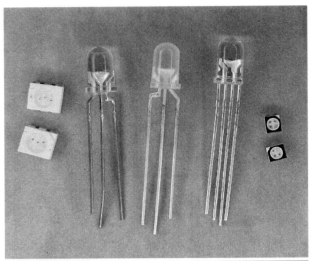

Figure 2-2 Photograph of bicolor and RGB LEDs

created. On the other hand, RGB LEDs allow more composite color possibilities through controlling the intensity of the three primary colors. The intensity control of LEDs is discussed in detail in the next section. LEDs are also used to make complex displays, such as seven-segment, alphanumeric, bar, and dot matrix displays, which are shown in the next chapter.

Controlling LEDs

Controlling LEDs refers to the ability of an electronic circuit to turn an LED on or off. If you wire up a circuit as shown in a previous illustration and set the DC power supply voltage to 5V, then the only way to turn the LED off is to turn the power supply off. However, LEDs can be turned on or off using the pins of a microcontroller, and interesting lighting patterns can be created. The next illustration shows a circuit diagram using a Tiny13 microcontroller and five LEDs.

Each of the LEDs is connected to a pin of the microcontroller with a series resistor. However, two of the LEDs (LED1 and LED2) are connected in the so-called current sink mode and the other three (LED3, LED4, and LED5) in the current source mode. These two different arrangements are used for the purpose of illustration only. The 74 series of TTL gates (e.g., 7400 or 74LS00) had more current sink capability than current source capability. So when using those gates, it was common to connect external loads (such as the LED) in current sink mode than in source mode. However, modern CMOS devices, such as the AVR series of microcontrollers, have symmetrical output current driving capability and, therefore, it does not matter whether the LEDs are connected in current sink or source mode.

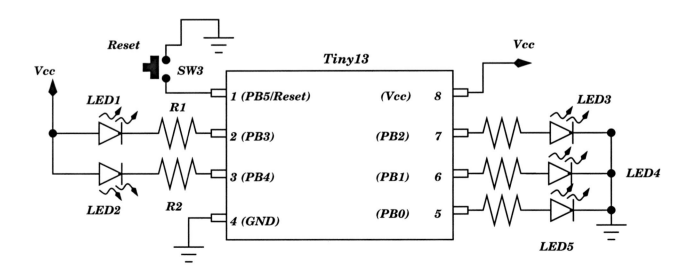

Coming back to the circuit shown in the previous illustration, LED1 and LED2 will light up when the respective output pin is set to logic "0," while the rest of the LEDs will light up when the respective output pin is set to logic "1." The value of the series resistor depends upon the amount of current desired through the LED, but the value should be no more than the output current capability of the microcontroller. AVR microcontrollers are capable of driving up to 40mA of current through each pin. The supply voltage of the system should be chosen such that the LED turn-on voltage is less than the supply voltage (Vcc). As an example, choosing 3V as Vcc (using two alkaline batteries) would be fine if you plan to use red LEDs in the circuit. However, blue LEDs won't work in such a situation. If you plan to use blue LEDs, the supply voltage Vcc should be 5V.

The microcontroller can turn each LED on or off at any rate. However, if the rate exceeds 20 Hz, then the LEDs would appear to be constantly turned on with an irritating flicker. On the other hand, if the rate is increased to, say, 100 Hz, the flicker would disappear and the LEDs would appear to be continuously on. The LEDs are indeed being turned on and off at a rate of 100 Hz, but the human eye is unable to follow that high a rate of change. This is an interesting phenomenon, and we use it to change the intensity of the LED, also called light intensity modulation.

The image on the bottom of this page illustrates a pulse width modulated (PWM) signal.

The signal has a frequency that is set to a high value. This frequency is $F = 1/T$, as shown in the top trace of the figure. Say F is set to 100 Hz. While keeping the frequency of the signal constant, the on time of the signal (i.e., the time for which the signal is at logic "1") is changed. Let T1 be the on time. The ratio of the on time (T1) to total time period (T) is called the duty cycle of the signal. Signals shown in the following illustration can easily be generated by the microcontroller circuit in the illustration on page 32. If the signal with 50% label is applied to LED3 through pin PB2 of the circuit, then the observer will perceive an intensity of 50%, compared to the case when the output at pin PB2 is permanently set to logic "1" without any change ever (when the pin PB2 is set permanently to logic "1," the LED will work at maximum intensity). This is because the average current through the LED is now 50%. Similarly, if

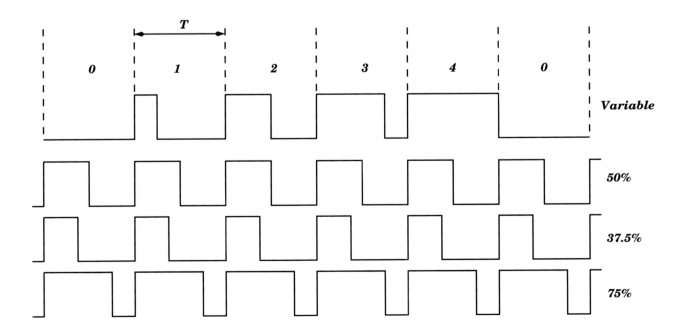

the signal labeled 75% is applied to LED3, its intensity will be 75% of the maximum intensity. The duty cycle of the signal can be set to any arbitrary value between 0% (minimum intensity) and 100% (maximum intensity). The PWM signal can be generated either using program control or through built-in hardware timers in the AVR microcontroller. Using the hardware timer allows the microcontroller to perform other additional tasks. Use of PWM software as well as hardware for LED light intensity control is illustrated in many subsequent projects. The use of PWM-based LED light intensity control is best illustrated in projects that use multicolor LEDs (such as RGB LEDs), where individual LED intensity control is used to create a large number of intermediate colors and shades.

Apart from intensity control, it is also pertinent to discuss the various ways in which LEDs can be connected. Until now, we have considered only a single LED per pin. But sometimes it may be required to connect multiple LEDs on a single pin of the microcontroller. Multiple LEDs can be connected together in two combinations: series or parallel. Connecting LEDs in series with a single resistor, as shown in this illustration, is possible.

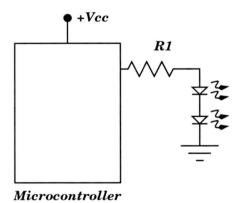

Microcontroller

The number of LEDs that can be connected in series and connected to the microcontroller pin will be determined by the LED turn-on voltage and the supply voltage. For a Vcc of +5V, two red LEDs can easily be connected in series. But two

blue LEDs cannot be connected, since the turn-on voltage of two blue LEDs in series is more than the +5V supply voltage. Similarly, three red LEDs cannot be connected in series to a +5V supply for precisely the same reason.

In case there is a need to connect three LEDs, a better configuration would be to connect the LEDs in parallel, as shown here:

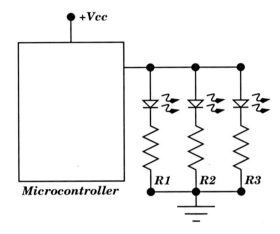

Microcontroller

Note that instead of using a single series resistor and connecting the LEDs in parallel, we have chosen to use one series resistor per LED and connect the LED-resistor configuration in parallel. This is because different LEDs of the same color and batch may have marginally different turn-on voltages and if a single resistor is used for the entire parallel configuration of the LEDs, the LED with the lowest turn-on voltage would dominate the other LEDs; hence, it is consuming more current and would thus appear to be brighter than the others. In a more extreme condition, this particular LED would hog all the current and may not allow other LEDs to even turn on. When connecting several LEDs in parallel, the sum of the current through all the LEDs should be less than the current source (or sink) capability of the microcontroller pin. In case the current through the LEDs exceeds the capability of the microcontroller, the configuration shown in the previous illustration (or the one that follows) should be used.

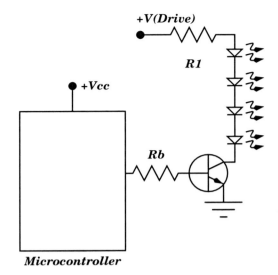

Microcontroller

The configuration in this illustration uses an NPN transistor to drive many LEDs arranged in series. The drive voltage for the LEDs—V(Drive)—must be more than the sum of the turn-on voltages of all the LEDs in series. The resistor R1 sets the current through the LEDs. The NPN transistor requires a base resistor (Rb) to limit the base current, and the value of Rb is calculated based on the collector current through the transistor (which is also the current through the LEDs) and the current gain of the transistor. Let's take an example: Assume that you want to connect five red LEDs in series and drive 30mA through them. From Table 2-1, the red LED turn-on voltage is 2V, so 10V will be required to forward-bias the LEDs. The transistor will have a drop of 0.5V across the collector and emitter terminals V(ce). A V(Drive) voltage of 15V would be desirable, and thus the value of the resistor R1 = (15 − 10.5)V/30mA = 150 Ohms. A suitable low-power transistor such as the BC547 would be suitable for this application. Typical β for BC547 is 100; therefore, the base current required would be 30mA/100 = 300μA. If the microcontroller is powered with a supply voltage of +5V, then the logic "1" voltage of 4.5V can be reasonably assumed. The V(be) drop across the base-emitter junction of the transistor is about 0.7V. Thus, Rb = (4.5 − 0.7)V/300μA = 12.6KΩ. So, a 10KΩ resistor for Rb is quite appropriate. For the configuration shown in the previous illustration, it is not necessary that all the LEDs in series be the same color. But in the calculation, the sum of forward voltage drop of all these LEDs must be taken into account to decide the V(Drive) voltage and thus the values of R1 and Rb.

The next illustration shows how multiple LEDs can be connected in parallel using an NPN driver transistor. This configuration would be desirable in case the sum of the currents through all the LEDs is larger than the microcontroller pin can supply. For example, suppose you want to drive 10 LEDs in parallel, each with 20mA current. The current requirement is 200mA, which is much more than a single pin of the microcontroller can supply. However, a medium-power NPN transistor with maximum collector current (Ic(max)) of, say, 1A would be suitable to drive these LEDs. The calculations required for the value of the series resistor for each of the LEDs, as well as the base resistor Rb, would be as shown earlier.

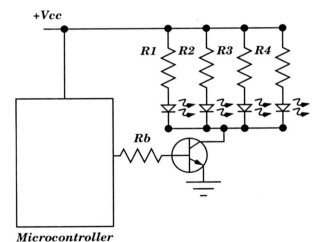

Microcontroller

Project 2
Flickering LED Candle

Even with all sorts of modern lighting methods, candles still capture the human imagination. A candle-lit dinner is considered more romantic than one with (even dim) normal lighting. Perhaps it's the way a candle flickers that makes it unique and worthwhile to emulate. In this project we show how an LED can be used to mimic candlelight. The next illustration shows the block diagram of the flickering LED candle. Lighting up the LED is not a problem. The trick to mimicking a candle lies in re-creating the way a candle flickers. The candle flame sways randomly (or so it seems), and sometimes the intensity of the flame seems to vary with air currents. When using an LED to behave like a candle, it may not be possible to make the "flame" sway, but what can certainly be achieved is the random variation of intensity, even in the absence of air currents. The block diagram shows a random number generator providing input to an intensity control circuit for the LED.

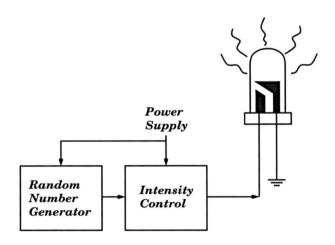

Randomization has always been one of the most confusing aspects of the implementation process. There has been endless talk about whether any number or noise generator can be truly random. The answer is simply "no." This is because every "random" number generator repeats itself after some interval. If this periodicity is sufficiently large, the source, which is not really random, appears to be completely random. Hence, we call such sources pseudo-random number generators. Embedded systems generally use some hardware-based pseudo-random number generators to introduce nondeterministic behavior in their working. However, a particular type of pseudo-random generators known as linear feedback shift registers (LFSRs) can be integrated easily in software alone, without the requirement of additional hardware resources.

The illustration at the bottom of the page shows the block diagram of an LFSR-based pseudo-random number generator. An LFSR is a shift register, the input to which is the exclusive-or (XOR) of some of the bits of the register itself. The bit positions that are used for the XOR operation to yield the input bit are called taps. The LFSR must be initialized with a non-zero seed value. If the seed is zero, LFSR would never leave its initial state because XOR of any number of zero-valued bits is zero. LFSRs have an interesting property: if the feedback taps are chosen carefully, outputs cycle through $2^n - 1$ sequences for an n-bit LFSR. The sequence then repeats after $2^n - 1$ instances. If the output sequences are observed, they appear to be random. A ten-bit LFSR is shown in this image. It will have a sequence length of 1,023. Similarly, a 16-bit LFSR would have a length of 65,535, and so on.

The LFSR described previously is called the Fibonacci LFSR. There is one more type of

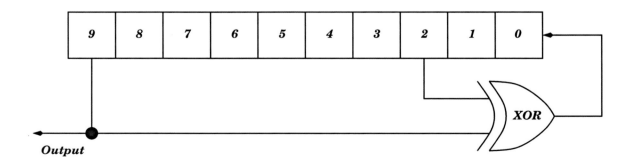

LFSR, known as the Galois LFSR, in which bits that are not taps are shifted unchanged. The taps, on the other hand, are XOR'd with the output bit before they are shifted to the next position. This project implements a Galois LFSR, while the implementation of a Fibonacci LFSR is shown in Project 4, later in this chapter.

Design Specifications

The aim is to develop a battery-operated flickering LED candle that mimics a real candle as closely as possible. The intensity of the LED can be varied using a pseudo-random number generator implemented using a suitable-length LFSR. The size of the LFSR would determine how soon the lighting pattern starts repeating. The pseudo-random number generator and the intensity control of the LED are to be implemented using one of the tinyAVR microcontrollers with the smallest pin-count. This illustration shows the final block diagram of the implementation.

The pseudo-random number generator is implemented by the tinyAVR microcontroller, and the port pins of the microcontroller implement the intensity control. To have light of a color similar to that of a candle, a white or warm white LED is suitable. However, this means that a supply voltage of 5V or more would be required. AVR microcontrollers operate up to 5.5V supply voltage, and this can be met easily with four AA 1.5V primary cells (such as alkaline) or rechargeable batteries such as NiMH, which have a terminal voltage of 1.2V.

The circuit, the LED, and batteries should also be packaged to look like a candle. With these specifications, let's see how the project is implemented.

Design Description

Figure 2-3 shows the schematic diagram of the tinyAVR-controlled flickering LED candle. We have used the Tiny13 controller, which is an eight-pin device. Figure 2-4 shows the schematic of the LED holder board. The circuit has been split into two boards: the controller board and the LED holder board. The idea was to arrange the LED holder board on top of the controller board so as to minimize the footprint of the circuit board. This way, it occupies less space and can be packaged easily in a tube to give the feel of a candle.

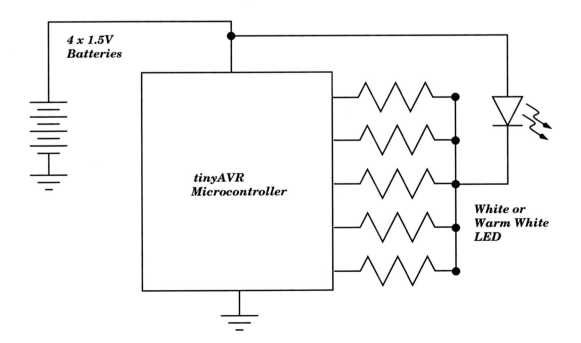

4 x 1.5V Batteries

tinyAVR Microcontroller

White or Warm White LED

Make sure that SL1,2,3 match with their respective SL1,2,3 on the LEDholder board

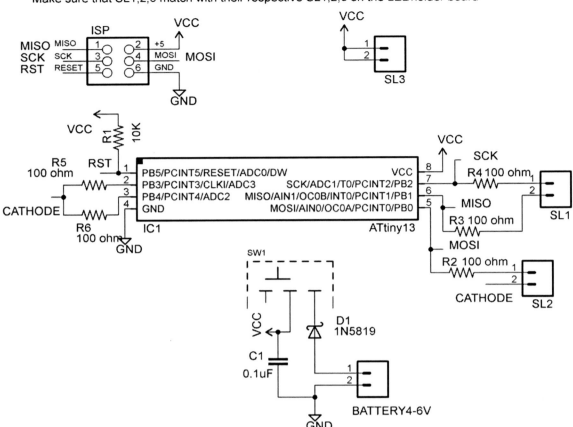

Figure 2-3 Flickering candle controller board: Schematic diagram

Figure 2-3 shows connectors SL1, SL2, and SL3, which are used to connect the LED on the LED holder board. The LED is arranged in the current sink mode and is placed on the second board, as shown in the schematic in Figure 2-4. The eight-pin controller has five I/O pins, and all are used to connect to the LED with a series resistance of 100 Ohms. The turn-on voltage of a white LED is 3.5V, so assuming a 5.5V supply voltage, each pin will sink around 20mA current, which can be handled easily by an AVR controller pin. Since five pins are used, the maximum current through the LED is 100mA. We chose to use a 1W high-brightness warm white LED, which can handle a maximum of 300mA current. The controller board also has an ISP connector to program the tinyAVR microcontroller and another

connector to connect the batteries. Switch SW1 is used to turn the power on or off to the circuit.

Figure 2-4 shows the LED holder board. It shows three connectors, and each of these

Make sure that SL1,2,3 match with their respective SL1,2,3 on the flickering candle board

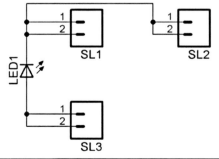

Figure 2-4 LED holder board: Schematic diagram

connectors mates with a corresponding connector on the controller board. Thus, SL1 on the LED holder board mates with the SL1 connector on the controller board and so on. LED1 is a 1W white high-power LED.

Fabrication

The layouts of both the boards in EAGLE, along with the schematic diagrams, can be downloaded from www.avrgenius.com/tinyavr1.

The controller board is mainly routed in the solder layer with few jumpers in the component layer. On the other hand, the LED holder board is routed in the component layer because the connectors have to be connected to the other side for proper mating with the controller board. Figures 2-5 through 2-9 show photographs of the project in various stages. Both the circuits are made on single-sided board. The tinyAVR chosen for the controller board is an SMD version, as are the resistors and capacitor.

Figure 2-6 The LED controller board. The Tiny13 controller is soldered on the solder side of the PCB. Notice the three jumper wires.

Figure 2-7 Track side of the LED controller board

Figure 2-5 The LED holder board mounted on the controller board

Figure 2-8 The LED holder board with a 1W white LED covered with hot glue drawn in a wick

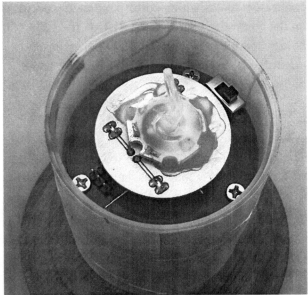

Figure 2-9 The completed flickering candle installed inside a Perspex tube. The battery holder for four AA alkaline batteries is below the LED controller board.

Figure 2-8 shows the photograph of the LED holder board. High-power LEDs, such as 1W LEDs, are usually available with heat sinks. Jumper wires were soldered to the pins of the LED and then soldered onto the PCB. After soldering the LED, a generous amount of hot glue melt was poured on the LED. While the glue cooled, it was gently drawn out to form a wicklike structure, as seen in Figure 2-8. A video of the flickering LED candle is available on our website.

Design Code

The compiled source code, along with the MAKEFILE, can be downloaded from www.avrgenius.com/tinyavr1.

The code runs at a clock frequency of 1.2 MHz. The controller is programmed using STK500 in ISP programming mode. During programming, the clock frequency is set to 1.2 MHz by selecting the oscillator frequency of 9.6 MHz and programming the CKDIV8 fuse, which divides the clock by 8. The control software for the flickering LED candle is quite simple and straightforward. The random

number generator is a 32-bit Galois LFSR with taps at 32, 31, 29, and 1 if the rightmost bit is numbered 1. Based on the random values generated by the LFSR, random numbers of LED channels are switched on. Between two port updates, a random amount of delay is executed. The value of the delay time is also derived from the LFSR value. The seed value of the LFSR is set to 1. The complete source code is given here:

```
#include<avr/io.h>
#define F_CPU 1200000UL
#include<util/delay.h>
int main(void)
{
  unsigned long lfsr = 1;
  unsigned char temp;
  DDRB= 0xff;
  while(1)
  {
    lfsr = (lfsr >> 1) ^ (-(lfsr & 1u) &
    0xd0000001u); /* taps 32 31 29 1 */
    temp = (unsigned char) lfsr;
        //take lowermost eight bits
    DDRB = ~temp; //Declare those pins as
```

```
   //output where temp has zero
   PORTB = temp; //Give the value of 0
   //to the pins declared output
   temp = (unsigned char) (lfsr >> 24);
   _delay_loop_2(temp<<7);
 }
}
```

The variable **lfsr** implements the actual LFSR. The variable **temp** takes the lowermost eight bits of the LFSR and switches on the random number of current sinking channels. It further takes the uppermost eight bits and provides random delays between the two port updates.

Project 3
RGB LED Color Mixer

It is an established truth that all the colors one sees are combinations of three primary colors—red, blue, and green. We developed this project to show how you can mix the primary colors in differing proportions and thus make millions of colors. Some of you might have already tested this hypothesis by defining custom colors in the color panels of popular PC graphics designing software like Microsoft Paint, Adobe Photoshop, etc. The displays of your PCs, laptops, or netbooks are also specified by the number of colors they support. Some support 15-bit colors, using 5 bits for each primary color, which implies that they can make 2^5 combinations to make a total of $2^5 \times 2^5 \times 2^5$ colors. Advanced displays support 24-bit colors or even more. In this project, we demonstrate this concept of color mixing on a single RGB LED. The user software generates eight-bit colors. So we can display $2^8 \times 2^8 \times 2^8$ colors on a single LED, but such a large number of colors cannot be resolved by the human eye.

Design Specifications

The aim is to develop an RGB LED-based design that allows us to mix variable percentages of red, green, and blue colors (in the form of LED lights here) to synthesize a plethora of colors. The percentage of each color should be configurable by the user. The technique used to control the intensity of each LED (analogous to varying the percentage of each color) is pulse width modulation, which has been described earlier. The block diagram is shown here:

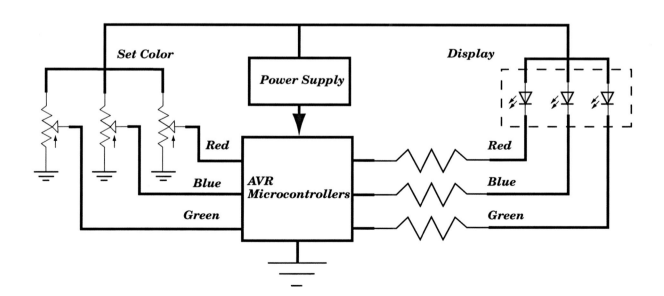

Design Description

Figure 2-10 shows the schematic diagram of the RGB color mixer project. The input voltage can vary from around 5.5V to 20V for the voltage regulator LM2940 to give a constant output voltage of 5V. Diode D1 is a Schottky diode (IN5819) with a forward voltage drop of approximately 0.2V. It protects the circuit in case the input voltage is applied with reverse polarity. Capacitors C2 and C8 are used to filter out the spikes and unwanted noise in the power supply. C4 and C7 are used to stabilize the output of LM2940. C3 is soldered near the supply pins of the microcontroller to further decouple the noise arising in the circuit. The heart of the project is the ATtiny13 microcontroller. It has all the necessary features like timers, ADC, I/O pins, etc., required for the project. The code discussed in the next section is small enough to fit into the 1KB Flash memory of this controller. The LED used is a common-anode RGB LED in a through-hole

package, and is connected to the microcontroller in current sink mode. Resistors R1, R2, and R3 act as current-limiting resistors for red, blue, and green LEDs, respectively, and are 100 Ohms each. SL2, SL3, and SL4 are the three potentiometers used for deciding the intensity level of each color. Capacitors C1, C5, and C6 are used to filter noise across the output of potentiometers. The outputs of all the potentiometers go to the ADC input channels of the microcontroller.

The circuit can also be designed without the use of any voltage regulator, but then the applied input voltage should be between 4.5V and 6V. The circuit diagram without the regulator is shown in Figure 2-11. It is identical to Figure 2-10, except the voltage regulator LM2940 and capacitors required at its output are absent.

The source code of the project reads the values of the potentiometers by using three channels of the ATtiny13's analog-to-digital converter and reflects the corresponding change in intensity of

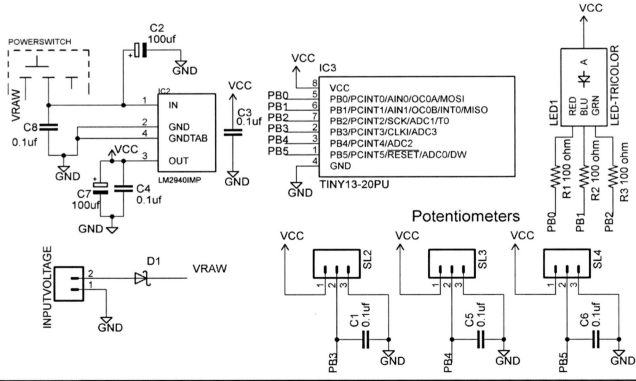

Figure 2-10 Color mixer: Schematic diagram

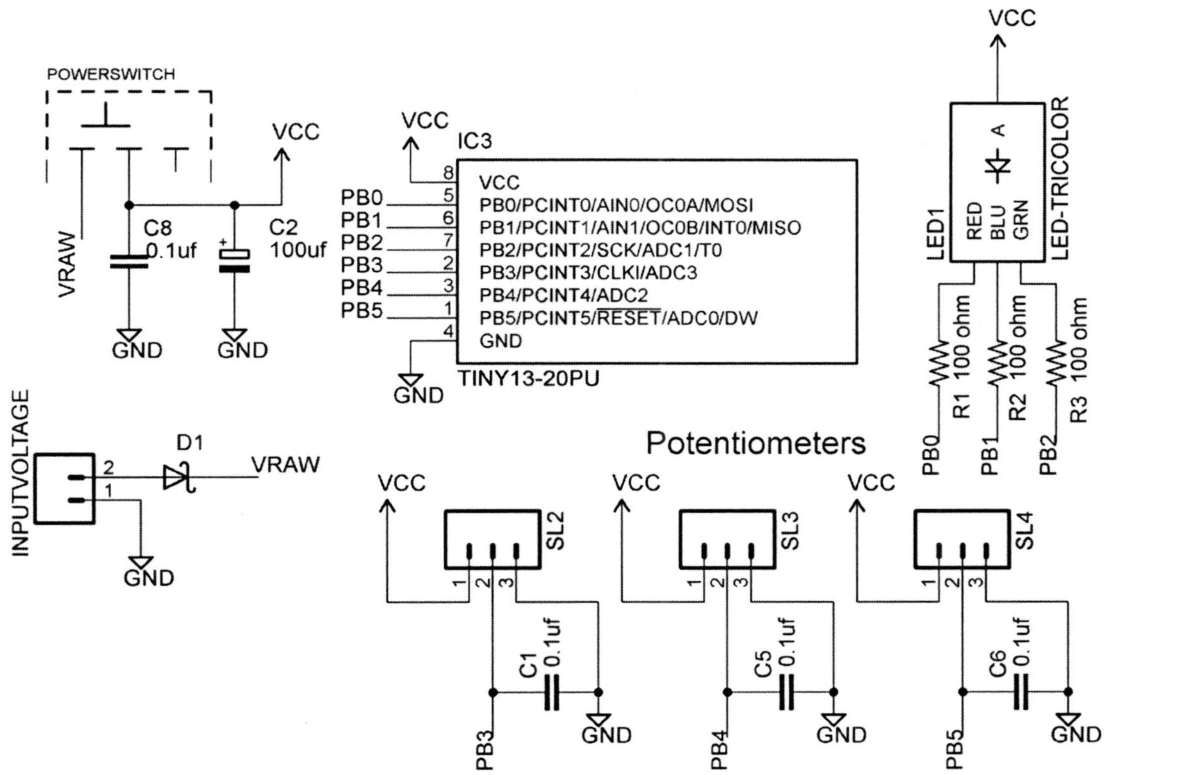

Figure 2-11 Color mixer: Schematic diagram without the regulator

three LEDs using a software-generated eight-bit (256 levels) pulse width modulation (PWM). By "software-generated PWM," we mean that the hardware PWM channels available on the ATtiny13 have not been used. ATtiny13 offers only two hardware PWM channels, but we require three channels to control three LEDs; therefore, a software-based approach has been adopted. tinyAVRs have ADCs with a ten-bit resolution, but only eight-bit resolution has been utilized in this project. Each ADC channel converts the analog voltage output from a potentiometer into a digital value lying within the range of 0 to 255 (eight-bit resolution), which can be mapped directly to the intensity level of the corresponding LED.

Fabrication

The board layouts of both versions, along with the schematic diagrams, can be downloaded from www.avrgenius.com/tinyavr1.

Both the boards are mainly routed in the solder layer with few jumpers in the component layer. The circle drawn in the layouts is 40 mm in diameter and used to make space for the standard ping pong ball that we have used to cover the RGB LED for proper diffusion and mixing of the colors. Figures 2-12 and 2-13 show the component and solder sides of the soldered board (with regulator), respectively. Figure 2-14 shows the component side with the LED covered by a ping pong ball.

Design Code

The compiled source code can be downloaded from www.avrgenius.com/tinyavr1.

The code runs at a clock frequency of 9.6 MHz. We have used the RESET pin of ATtiny13 as an ADC input channel, so the reset functionality of this pin has to be disabled by programming the RSTDISBL fuse bit. As soon as this fuse bit is programmed, the ISP interface becomes

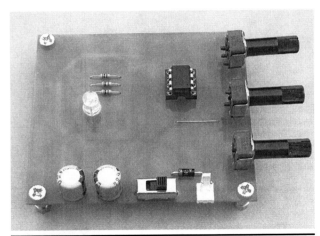

Figure 2-12 Color mixer: Component layout

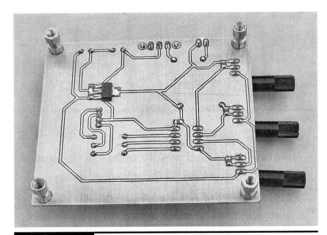

Figure 2-13 Color mixer: Solder side

Figure 2-14 Color mixer: Component side with the LED covered

unavailable and the controller has to be further programmed using a different interface. Hence, the controller is programmed using STK500 in HVSP programming mode. The important sections of the code are explained here:

```
//Overflow routine for Timer 0
ISR(TIM0_OVF_vect)
{
  //Value of e decides the no of levels
  //of PWM.
  //This has 256 levels for e - 0 to 255
  //0 - completely on and 255 is
  //completely off
  if(e==255)
  {
    e=0;
    PORTB |= (1<<0)|(1<<1)|(1<<2);
  }
  abc(pwm[0],pwm[1],pwm[2],e);
  e++;
}
```

This piece of code is the Timer0 overflow interrupt subroutine. This routine handles the three software channels of PWM and is called whenever Timer0 overflows. The variable **e** maintains the state of the PWM cycle. When the value of e becomes "255," all the LEDs are switched off and e is reinitialized to "0." The array **pwm** specifies the value of the duty cycle of each LED. The code has been written in such a way that the value "0" in any variable of array **pwm** corresponds to maximum duty cycle (LED completely on) and "255" corresponds to minimum duty cycle (LED completely off).The function **abc** compares each value of array **pwm** with **e** and if a match occurs, it switches on the corresponding LED.

```
//This function reads the value of ADC
//from selected channel
unsigned char read_adc(unsigned char
    channel)
{
  unsigned char k;
  unsigned int adcvalue=0;
  ADMUX = ADMUX&(0b11111100);
```

```
    //clear channel select bits
ADMUX |= channel;
    //neglect first reading after changing
    //channel
ADCSRA |= 1<<ADSC;
while(ADCSRA&(1<<ADSC));//Wait
adcvalue=ADCH;
adcvalue=0;//neglect reading
for(k=0;k<=7;k++)
{
  ADCSRA |= 1<<ADSC;
  while(ADCSRA&(1<<ADSC));//Wait
  adcvalue += ADCH;
}
return (adcvalue>>3); //divide by 8
}
```

This subroutine handles the conversion of the analog input on the selected ADC channel (passed to this function through the argument **channel**) into a digital value between 0 and 255, and returns this value to the calling function. The function first selects the required ADC channel. It neglects the first ADC reading and returns the average of the next eight readings to the calling function. The ADC is used in the single conversion mode here. After selecting a new ADC channel, it is always recommended to ignore the first reading in order to avoid some conversion errors.

The **main** function runs in an infinite loop. It takes the readings of the three ADC channels, one by one, and assigns them to the corresponding variables of the **pwm** array.

Working

The intensity of each LED can be varied from minimum to maximum in 256 continuous levels by using the potentiometers. Different intensities of red, green, and blue LEDs produce different colors. The ping pong ball ensures proper mixing of the three components.

Project 4
Random Color and Music Generator

We successfully incorporated an LFSR-based pseudo-random number generator in a Tiny device in Project 2, but the seed of the LFSR was fixed. This means that each time you switch on the board, it generates the same pattern periodically. In this project, we show how a 16-bit LFSR can be integrated in a device as small as an ATtiny13 with a floating ADC channel used as its initial seed. Using a floating ADC channel as a seed gives different initial values to the LFSR each time the circuit is switched on, and the patterns appear even more random. Although a 16-bit LFSR can ideally generate random numbers (excluding 0) with a periodicity of 65,535, as explained before, this requires the taps to be at specific locations. An n-bit LFSR with a period of $2^n - 1$ is known as a maximal LFSR, which is what this project uses. Furthermore, the randomization in the output of the LFSR is demonstrated by producing various colors on an RGB LED and sounds on a speaker.

Design Specifications

The aim is to integrate an LFSR in a Tiny device and demonstrate its random behavior on an RGB LED and a speaker. The circuit should work at input voltages as low as 3V. The technique used in controlling the intensity of each LED to generate random colors is again pulse width modulation, as used in Project 3. Random sound is generated by using square waves of different audible frequencies to drive a speaker. The following illustration shows the block diagram of the project.

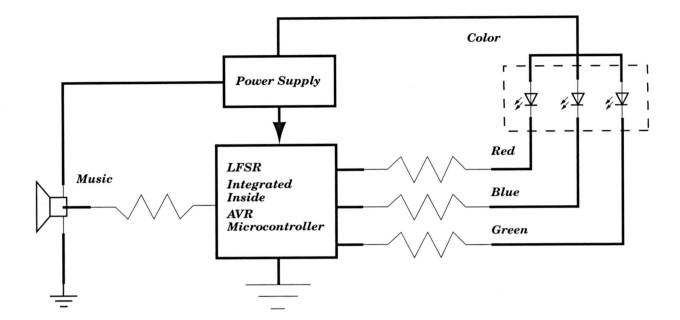

Design Description

Figure 2-15 shows the schematic diagram of the project. MAX756 is a step up dc-dc converter operated in the 5V mode here. It generates 5V for the circuit from an input source of voltages as low as 3V. If the input voltage exceeds the desired output voltage (5V), the output voltage follows the input voltage. This can damage the other parts of the circuit, including the MAX756, so it is required that the input voltage never be allowed to exceed 5V. Diode D1 and inductor L1 are required for the operation of MAX756. The controller used is again ATtiny13, having all the necessary features required for the project. The LED used is a common-anode RGB LED in SMD package, connected to the microcontroller in current sink mode. Resistors R5, R6, and R7 act as current-limiting resistors for Red, Blue, and Green LEDs respectively and are of 100 Ohms each. Part T1 (2SD789) is an NPN transistor which acts as the driver for the speaker. It has to be used because the I/O pins of the AVR can only source or sink 40mA current, which is not sufficient for the speaker. The speaker used is rated at 0.5W and has a resistance of 8 Ohms. In order to keep the power dissipated by the speaker within permissible limits, it should be operated with a 10 Ohm resistor in series. PB4 pin of the controller is used as a floating ADC channel to get the initial seed for the linear feedback shift register (LFSR).

The project generates pseudo-random numbers using a 16-bit Fibonacci LFSR. The numbers thus generated are used to change the color of the RGB LED and the tone of the speaker after every half a second. The colors have been created using a software-generated PWM of 10 levels, as opposed to a 256-level PWM in Project 2. This means that the LED can show $10 \times 10 \times 10$ colors, but only 16 of these have been selected for display. The tone for the speaker is generated by applying a square wave of varying frequency. A total of nine different frequencies or sounds have been selected.

Fabrication

The board layout, along with the schematic, can be downloaded from www.avrgenius.com/tinyavr1.

The board is routed in the solder layer with few jumpers in the component layer. The component layout is shown in Figure 2-16, and the solder side is shown in Figure 2-17. In the soldered board, the

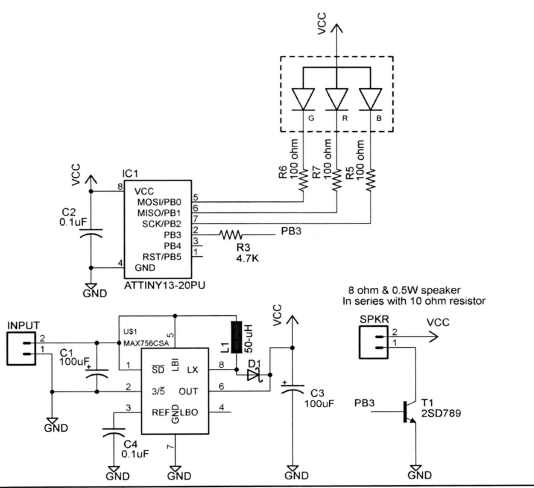

Figure 2-15 Random color and music generator: Schematic diagram

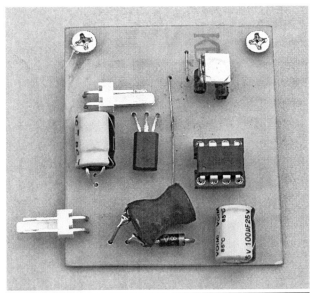

Figure 2-16 Random color and music generator: Component layout

Figure 2-17 Random color and music generator: Solder side

SMD RGB LED has been brought towards the component side by using berg strips.

Design Code

The compiled source code, along with the MAKEFILE, can be downloaded from www.avrgenius.com/tinyavr1.

The code runs at a clock frequency of 9.6 MHz. The controller is programmed using STK500 in ISP programming mode. The important sections of the code are explained here:

```
while(1)
{
 //Wait for the seed from ADC
 if(i==4)
 {
  //This is the software code for LFSR
  bit = (reg & 0x0001) ^((reg & 0x0004)
   >> 2) ^((reg & 0x0008) >> 3)^((reg&
  0x0020) >> 5);
  reg = (reg >> 1) | (bit << 15);
 //Sound-generating code
  PORTB |= 1<<3;
  delay(t);
  PORTB &= ~(1<<3);
  delay(t);
 }
}
```

This is the main infinite loop of the program. It starts only when the value of i is 4, which means that it takes the fourth value of the floating ADC as its initial seed. The ADC interrupt executes four times and increments the value of i by 1 each time. When the value of i reaches 4, the value of the floating ADC channel is assigned to the variable **reg** and the ADC interrupt is then disabled. The variable **reg** is a 16-bit integer that implements the actual LFSR. The taps for the LFSR have been taken at bit positions 16, 14, 13, and 11 if the leftmost bit is numbered 1. The sound-generating code simply generates the square wave of the selected frequency.

```
//Timer0 overflow ISR
ISR(TIM0_OVF_vect)
{
 //for color
 if(e==9)
 {
  e=0;
  //Start of new cycle
  PORTB = PORTB|(1<<2)|(1<<1)|(1<<0);
 }
 abc();
 //Time to change
 //Approximately half a second
 j++;
 if(j==128)
 {
  a = reg%9;//get new value for sound
  t = pgm_read_word(d+a);
  a = reg%16;//get new value for color
  blue = pgm_read_byte(k+a);
  red = pgm_read_byte(l+a);
  green = pgm_read_byte(m+a);
  j=0;
 }
}
```

This function is the Timer0 overflow interrupt subroutine. It performs two main functions. First, it handles the software PWM just as in Project 2, and second, after every 128 overflows, it selects a new value of color and tone by taking the modulus of the **reg** variable. There are nine sounds stored in the array **d** (in the form of time delays for square wave), and taking modulo9 of **reg** assigns a random value from 0 to 8 to **a**. Then, the appropriate variable of array **d** is stored in **t**, which is used as the delay variable in the main infinite loop. Similarly, there are 16 colors and modulo16 of **reg** gives a random value from 0 to 15 to **a**. The corresponding intensity levels of each color from the arrays **k**, **l**, and **m** are stored in the corresponding variables **blue**, **red**, and **green**. Timer0 has been prescaled in such a way that 128 overflows occur in approximately half a second. The functions **pgm_read_byte** and **pgm_read_word** are used to fetch the constants stored in program memory instead of data memory. **pgm_read_byte** is used for 8-bit variables, and **pgm_read_word** is used for 16-bit variables. In order to save data memory, it is a good idea to store those variables that are constants (whose values don't change) in program memory by applying the attribute PROGMEM after their declaration. You can refer to the complete source code to obtain further details.

Project 5
LED Pen

You might have seen advertisements by battery manufacturers in which images are drawn in the air and captured by a long-exposure camera. You could do that with a flashlight and draw in the air and capture it on a camera. Now imagine, instead of a flashlight, you have a multicolor LED pen with which to draw these pictures. This is exactly what this project achieves. Some of the images drawn with an LED pen and captured with a camera are shown in Figure 2-18. Such images and pictures are called light doodles.

Figure 2-18 Photographs of some objects drawn in the air and captured with a long-exposure digital camera.

Design Specifications

An LED light pen is expected to be easy to hold in a hand so that one can draw and write easily, so size was a critical issue in this design. To achieve this objective, it was important to (a) use the smallest batteries as possible and (b) reduce the circuit complexity. The LED light pen is similar to the random color and music project, but the objective and implementation are totally different. This project shows two implementations of the LED light pen. We implemented the first version, and the block diagram is shown here:

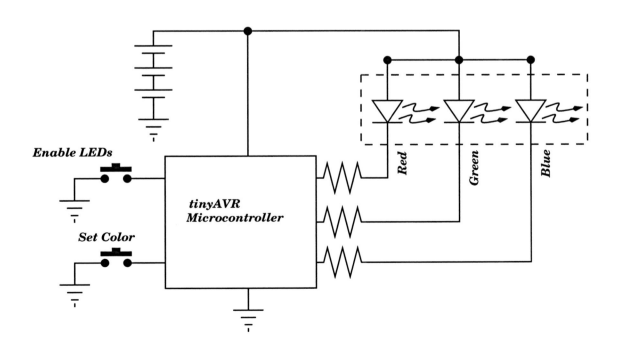

After we made this version and used it for a while, we realized the difficulties associated with it and made the second version shown at the bottom of this page.

Let's first discuss the requirements. The idea of a multicolor LED light pen came about after we saw light doodles on a website. The doodles were drawn using simple LED pens made with a few fixed-color LEDs. We thought, instead of using multiple LED pens, we could use a single pen with an RGB LED. The intensity of the light from the individual LEDs of the RGB LED could be controlled with a microcontroller to provide several colors, many more than are possible with individual fixed-color LEDs. To draw these images, you need a digital camera with an option to set the exposure and aperture manually. The maximum exposure time on the camera would determine how long you can draw with the LED pen. Typical compact cameras provide about a minute of exposure time, which is sufficient for simple light doodles. For more complex doodles, a single lens refractor (SLR) camera, which allows extremely long exposure time, is needed. Also,

when drawing the doodles, you might want to change the color of the light. A single switch could be used to select from a set of available colors through toggling. In addition, when you draw, you may want to turn the LED off while starting another segment of the image or picture; thus, a light on/off switch is also required. This is shown in the first block diagram of our LED pen. The pen has two switches: one to enable the LEDs and another to select the color.

The first version of the LED pen offered 16 different colors. The problem we faced with this version was that when you select a color and finish drawing and say you want the previous color, you have to step through all the colors to go back to that last color. This wastes precious time. We wanted a mechanism to quickly select the color rather than step through all the possibilities. Thus, we replaced the "set color" switch with a potentiometer, which can be rotated quickly to get the desired color. This scheme is shown in the second version. The rest of the features are the same as in version 1 of the LED pen. To achieve the objective of small size, we decided to use

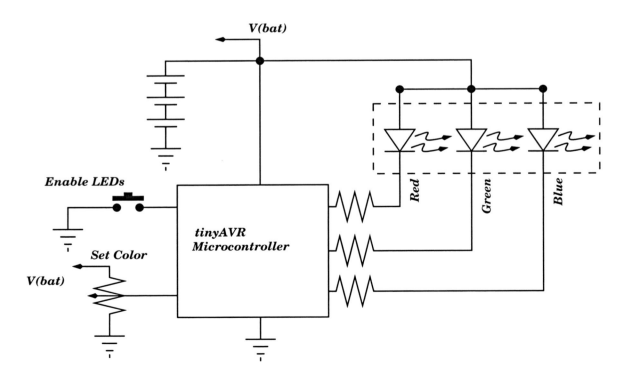

button cells to provide power to the circuit and to use an eight-pin tinyAVR microcontroller in DIP version with the help of a socket. Using a DIP version of the microcontroller allows us the freedom to remove the IC from the socket during the development of the pen, and thus we could avoid an on-circuit ISP connector to reduce the PCB size.

Design Description

Figure 2-19 shows the schematic diagram of version 1 of the LED pen. A Tiny13 microcontroller is used to read the two switches, S1 and S2, and an RGB LED is used to provide various colors. The circuit is powered with the help of three LR44 coin cells. The circuit does not have any power-on/-off switch. The microcontroller is programmed such that if the LED light is turned off (with the help of switch S2), the microcontroller enters a low-power operating mode—the power-down mode of operation— which reduces the current consumption. Typically,

an AVR microcontroller in power-down mode consumes only a few microamperes of current.

Figure 2-20 shows the schematic diagram of version 2 of the LED pen. Here, switch S2 is replaced with a potentiometer (labeled POT1). The AVR13 microcontroller has several channels of the ADC. The center tap of the potentiometer is connected to an ADC channel input. The other two terminals of the potentiometer are connected to the Vcc and ground signals of the circuit. The center tap of the potentiometer will provide a voltage between Vcc and ground to the ADC channel input. The microcontroller uses preset thresholds on the input voltage provided by the potentiometer to select the color by setting the duty cycle of the PWM signal for each of the individual red, green, and blue LEDs. The RGB LED, as represented by SL1 in the figure, is connected in current sink mode. For both versions of the LED pen, the connector SL1 refers to a small circuit board with an RGB LED, as shown in Figure 2-21.

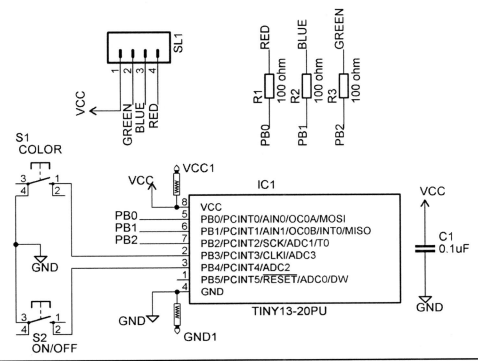

Figure 2-19 LED pen version 1: Schematic diagram

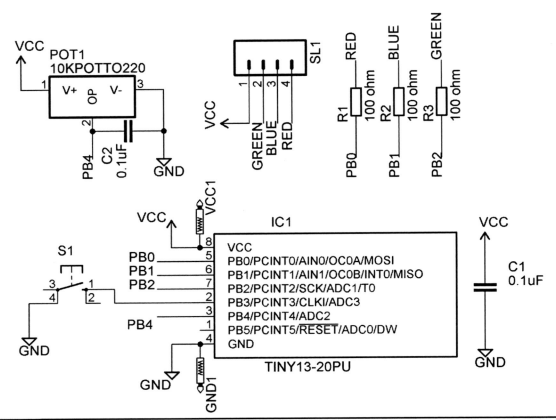

Figure 2-20 LED pen version 2: Schematic diagram

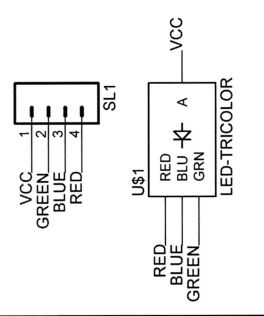

Figure 2-21 LED pen RGB LED board: Schematic diagram

Fabrication

The board layouts, along with the schematics, can be downloaded from www.avrgenius.com/tinyavr1.

Both versions of the LED pen are fabricated on a single-side PCB. The completed PCB is small enough to be inserted into a Perspex tube of 20 mm diameter. Figures 2-22 and 2-23 show the controller board of the LED pen version 2. Figure 2-24 shows the small PCB with the RGB LED mounted at a right angle to the main controller board. The circuit is powered with three LR44 coin cell batteries. Figure 2-22 shows the way the batteries are installed. The three batteries are held together with small magnets and the end batteries directly soldered to the PCB. The battery and magnet arrangement is glued together with rubber glue so that the batteries do not get disconnected. Figure 2-25 shows the photograph of version 1 of the LED pen.

Figure 2-22 Photograph of the top view of version 2 of the LED pen

Figure 2-23 Photograph of the bottom view of version 2 of the LED pen

Figure 2-24 The RGB LED PCB connected at a right angle to the main circuit board

Figure 2-25 Photograph of the side view of version 1 of the LED pen

Design Code

The compiled source codes for both the versions can be downloaded from www.avrgenius.com/tinyavr1.

Both systems run at a clock frequency of 1.2 MHz. The codes of both versions are similar, except that in version 1, the color change is performed through a switch, while in version 2, it is performed through a potentiometer. The programming of the code and fuse bytes is done using STK500 in ISP mode. Different colors are generated, again, by using software PWM channels for each color. Nine-level PWMs for each color have been selected. Out of the 9 × 9 × 9 colors possible, 16 have been selected and stored in the program memory. Also, there is one additional mode, called "run mode," in which all the colors appear one by one with a time gap of 500ms. The important code sections are summarized here:

```
ISR(TIM0_OVF_vect)
{
  DDRB &=~(1<<0|1<<1|1<<2);
  if(e==8)
  {
    e=0;
    xyz();
  }
  abc(pwm[0],pwm[1],pwm[2],e);
  DDRB |=(1<<0|1<<1|1<<2);
  e++;
  if(i==15)//Run mode
  {
    counter++;
    if(counter == 390)//500ms
    {
      counter = 0;
      if(k==14)
      k=0;
      else k++;
      pwm[0] = pgm_read_byte(&Blue[k]);
      pwm[1] = pgm_read_byte(&Red[k]);
      pwm[2] = pgm_read_byte(&Green[k]);
    }
  }
}
```

This section of code is again the Timer0 overflow interrupt subroutine and handles the nine-level software PWM, as explained in previous projects. Apart from this, if the selected mode is run mode, it scrolls through all the colors with a time gap of 500ms. This subroutine is common to both versions.

```
if(!(PINB&(1<<3)))
{
  _delay_ms(30);
  while(!(PINB&(1<<3))); //wait
  _delay_ms(30);
  TIMSK0 &= ~(1<<TOIE0);
    //Clear timer interrupt
  DDRB &=~(1<<0|1<<1|1<<2);
  GIFR |= 1<<PCIF;
    //Clear pending interrupt
  GIMSK |= 1<<PCIE;
    //Pin change interrupt enable
  MCUCR |= (1<<SE|1<<SM1);
    //Power down mode setting
  sleep_cpu();
    //CPU halted till interrupt
}
```

This segment runs in the main infinite loop when the device is in active mode. It checks the state of the switch on PB3. On press and release, it disables the Timer0 interrupts and also configures the I/O pins controlling the LED as floating. This switches off the LED completely and enables the pin-change interrupt. Although the pin-change interrupt can be executed on all I/O pins of the ATtiny13 controller, in the initialization part of the code, it has been set so that only PB3 pin can cause this interrupt. Then the setting in the **MCUCR** register selects the sleep mode as power down and sends the device into power-down mode by calling the function **sleep_cpu()**, from which it can only be woken up by the asynchronous pin change interrupt, which is explained subsequently. During the power-down mode, the code stops executing. It only restarts when there is a wake-up source, which, in this case, is the pin change interrupt on PB3.

```
ISR(PCINT0_vect)
{
  _delay_ms(30);
  while(!(PINB&(1<<3))); //wait
  _delay_ms(30);
  MCUCR = 0x00; //sleep disable
  GIMSK &= ~(1<<PCIE);
    //Pin change interrupt disable
  TIMSK0 = 1<<TOIE0;
    //Overflow Interrupt Enabled
}
```

This is the interrupt service routine for the pin change on PB3. The source code has been written in such a way that this interrupt is active only when the device is in power-down mode. This subroutine disables the sleep mode along with the pin-change interrupt. It also enables the timer again so that the generation of colors starts. This method of sending the controller into power-down mode is also common to both versions.

In the main infinite loop, the colors are selected by the switch in version 1 and by the ADC reading in version 2. Refer to the full source code for further details.

Conclusion

In this chapter, we have learned about various types of LEDs and their series and parallel control. We have also discussed in detail the intensity control of LEDs using PWM. These concepts have been augmented by four different projects. Apart from this, the LEDs can be arranged in multiple configurations in such a way that the ratio of the number of I/O pins required to the number of the LEDs controlled is less than unity, i.e., a given number of I/O pins can control a larger (than the number of I/O pins) number of LEDs. The PWM for all the projects, when required, was generated through software. tinyAVRs also have hardware PWM channels associated with timers. We discuss all this in the next chapter.

CHAPTER 3

Advanced LED Projects

IN THE LAST CHAPTER we had a glimpse of LEDs and how to use them in simple projects. The number of LEDs was small, and we showed a few projects using eight-pin tinyAVR microcontrollers. However, if we want to make projects with a large number of LEDs, then we need microcontrollers with more pins. Even with a large number of pins, many times LEDs outnumber them and conventional ways of connecting one LED to one microcontroller pin would not suffice. In this chapter we cover more LED-based projects demonstrating the advanced techniques of LED control. Some of the projects use single-color LEDs, while the rest use RGB LEDs in interesting applications. We use two schemes to control large number of LEDs with limited microcontroller pins: multiplexing and Charlieplexing. Charlieplexing is an extreme case of multiplexing and has become quite popular recently.

It is not necessary to connect the required number of LEDs on your own. A number of LEDs already connected in different colors, packages, shapes, and configurations are commercially available, as shown in Figure 3-1. The figure shows an 8 × 8 dual-color, dot matrix display (top-right corner), a 16 × 16 display (bottom-right corner), a seven-segment display (bottom-left corner), and two 5 × 7 dot matrix displays. The size of the display has an impact on the current rating of the LEDs; a larger display would also have LEDs that can handle larger current so that they are relatively brighter.

Figure 3-1 LED display packages

Multiplexing LEDs

Controlling multiple LEDs to create different patterns or display text is a common requirement. In the last chapter we showed how LEDs are controlled using microcontroller pins. With those techniques, we connected one LED to one pin of a microcontroller. Depending upon the way the LED was connected to the pin, the LED could be turned on or off by setting the logic on the pin to "1" or "0". The intensity of the light from the LED could also be controlled by generating a PWM signal on the microcontroller pin. However, using one microcontroller pin for each LED is wasteful. Instead, a technique called multiplexing is used. Multiplexing uses microcontroller pins on a time-sharing basis with multiple sets of LEDs. The following illustration shows the basic multiplexing scheme. In this figure, three rows and three columns are used to control nine LEDs. Each of the row and column pins is connected to one microcontroller pin. Thus, using six microcontroller pins, we are able to control nine LEDs. This scheme can be further scaled to a certain extent by increasing the rows and columns suitably. However, to maximize the utilization of the pins, it is advisable to keep the number of rows and columns equal as much as possible. To what extent can this scheme be scaled? Is it possible to control, say, 225 LEDs using a matrix of 15 rows and 15 columns? The answer lies in the peak current rating of the LEDs that are used for the display.

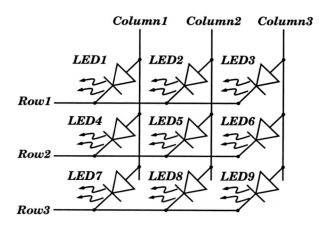

Let us first explain the operation of the multiplexed display as shown in the illustration. The LEDs are connected between rows and columns of wires. Each row and column, in turn, is connected to a pin of the microcontroller.

To turn a particular LED on, its column must be connected to Vcc and its row connected to ground. Suitable current-limiting resistors are assumed in the path of the current from Vcc to ground via the LED. Once a particular column is connected to Vcc (through the microcontroller pin), several LEDs on that column can be turned on by connecting the corresponding rows to ground (again, through the microcontroller). For example,

if Column1 is set to Vcc and Row1 and Row3 are set to ground, LED1 and LED7 would be turned on. Suppose LED1, LED2, and LED5 are to be turned on. Then Column1 (for LED1) and Column2 (for LED2 and LED5) would need to be connected to Vcc and Row1 (for LED1 and LED2) and Row2 (for LED5) would need to be connected to ground. However, if these rows and columns are set to the designated voltages, then LED4 would also turn on, since Column1 is at Vcc and Row2 is at ground! Thus, to avoid unintended LEDs from getting turned on, the voltages on rows and columns are activated in the following fashion.

For a time T, Column1 is set to Vcc and Column2 and Column3 to ground. In this time interval, Row1 is set to ground if LED1 needs to be turned on; otherwise, it is set to Vcc. Row2 is set to ground if LED4 needs to be turned on; otherwise, Row2 is set to Vcc; finally, Row3 is set to ground if LED7 needs to be turned on; otherwise, Row3 is set to Vcc. After the time interval T is over, another time interval T is started, and in this interval, Column1 is set to ground, Column2 is set to Vcc, and Column3 is set to ground, and Row1, Row2, and Row3 are set to Vcc or ground depending upon whether LED2, LED5, and LED8 need to be turned off or on. After this time interval, a third time interval starts, also for time period T, and in this period, Column1 and Column2 are set to ground and Column3 is set to Vcc. Row1, Row2, and Row3 are set to ground or Vcc depending upon whether LED3, LED6, and LED9 need to be turned on or off, respectively. After the end of the time period T, the cycle is repeated again.

What is the duration for the period T? This depends upon the number of columns. Each column must be turned on every 100 Hz or more. Thus, for a case with three columns, 3T = 10 ms (10 ms is the period of a 100-Hz waveform). Therefore, T = 3.3 ms.

The current-limiting resistor to be put in series can either be connected to the cathode, as shown in the bottom illustration, or to the anode, as shown in the subsequent illustration. However, the placement of the resistor decides the rating of the switches on the high (S1, S2, and S3) and low (S4, S5, and S6) sides. Switching the Vcc (high-side switching) requires a PNP transistor or a P-channel MOSFET. Switching ground (low-side switching) requires an NPN transistor or an N-channel MOSFET.

Let us consider the ratings of the high-side and low-side switches when the current-limiting resistors are placed in series with the LED cathode as shown in the illustration here.

In this scheme, S1 is turned on for a period T during which S2 and S3 are turned off. The current through the LED is determined by the Vcc supply voltage, the forward voltage drop across the LED and the series resistor, assuming negligible voltage drop across the low-side and high-side switches.

$$I(LED) = (Vcc - V(LED))/R$$

R is the series resistance. However, this current is the peak current. The average current through the LED is further reduced by a factor N, where N is the number of columns (in our case here, N = 3). To restore the current, the resistor value must be reduced by this factor N, thus increasing the peak current. However, the peak current cannot be arbitrarily scaled, since at some point, it would exceed the peak current rating of the LED. Typically, the peak current can be five to ten times more than the maximum average current. Therefore, the number of columns can be, at most, ten. If one is willing to reduce the required average current in the implementation to less than the maximum average current rating of the LED, then N can be increased further. Let us take an example to illustrate the issue. Table 2-1 in the previous chapter shows the electrical and optical data of LEDs. Red LEDs have a maximum average current of 30mA and a peak forward current of 120mA.

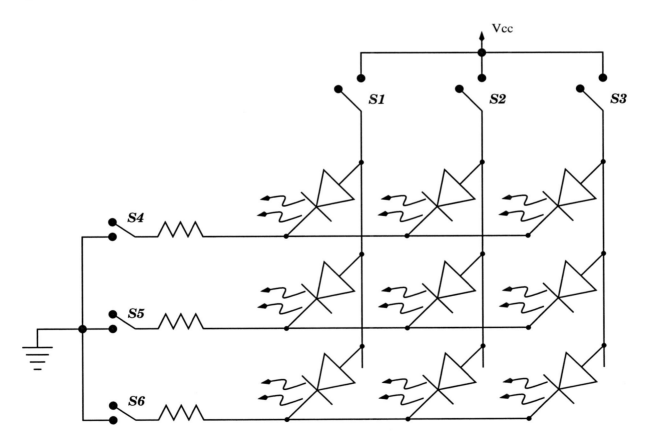

Thus, the value of R should be chosen such that peak current through the LED never exceeds 120mA. Now, if the number of columns is ten, the average current, due to multiplexing, will be 12mA, which is well within the maximum average current rating. We can increase the number of columns to 20 at the cost of reduced average current (and, therefore, the intensity of the LED light). The impact of putting the current-limiting resistor in series with the cathode, as shown in the previous illustration, is that the switches S4, S5, and S6 will have to handle a maximum current equal to the peak LED current, which for the present case is 120mA. However, S1, S2, and S3 would have to be rated to handle 360mA each. Typically, a NPN transistor or N-channel MOSFET can handle larger current than a corresponding PNP transistor or P-channel MOSFET. Instead of

the current-limiting resistor being put in series with the cathode, we can choose to put the resistor in series with the anode, as shown in the next illustration. However, this will require a corresponding change in the multiplexing scheme. Instead of connecting a column to Vcc, we need to connect a row to ground for time period T and, depending upon which LED we need to light up, switch the corresponding column to Vcc. In the next time period T, the next row is connected to ground and so on. The column switches (S1, S2, and S3) will now need to handle maximum current of 120mA, while the row switches will need to handle a maximum of 360mA. But low-side switches will be implemented using NPN transistors (or N-channel MOSFETs), and these are more easily available than their PNP or PMOS counterparts.

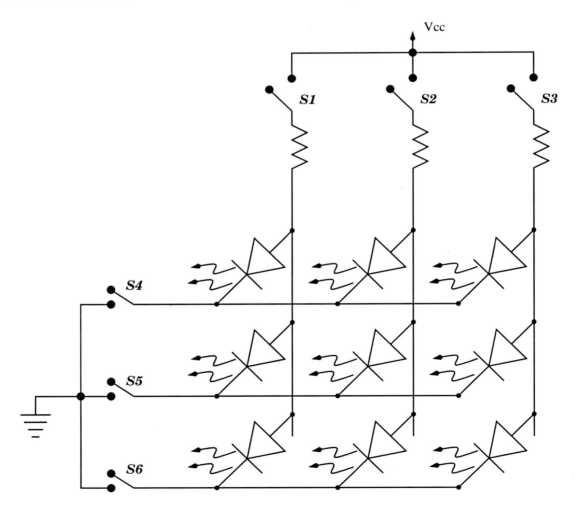

What if the number of LEDs is much larger than can be arranged in a single matrix of a number of rows and columns? The scalability of a matrix of LEDs is an issue, as discussed earlier, and the peak forward current of the LED doesn't allow one to scale the size of the display beyond a certain point. In such a situation, multiple matrices of LEDs controlled with independent rows and columns can be implemented as shown here:

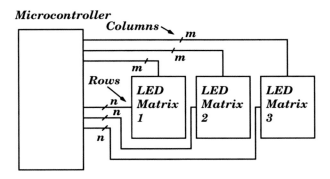

In such a situation, the availability of a sufficient number of microcontroller pins could be an issue. Rather than using a microcontroller with more pins, one could use external shift registers to increase the pins. The next illustration shows a scheme to increase the number of pins using a serial-in-parallel-out shift register such as the 74HC164 or the 74HC595.

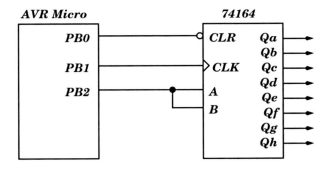

These two are eight-bit shift registers, that is, they provide eight bits of output data. Further, these shift registers could also be cascaded to provide 16 or 24 bits of data, as shown in the upcoming illustration. The shift registers provide only the pins required to connect a large number of LEDs (either directly to each pin or in a multiplexed fashion), but do not have current drive capabilities. Additional current boost transistors or MOSFETs are required.

The following illustration shows a scheme that uses 3 output pins from the microcontroller to get 16 output pins. However, it would take a program to shift the 16 bits of data onto the 16 output pins. The program would start by clearing the two shift registers (by generating a pulse on the PB0 pin), and then the required data is put on the input of the upper shift register. The lower shift register gets its data from the Qh signal of the upper shift register. After setting the required logic on the microcontroller pin connected to input of the shift register, the data is shifted in the shift register by generating a clock pulse on the clock input pin of the shift registers (PB1 pin from the microcontroller). Each clock pulse shifts the data from the input to Qa, from Qa to Qb, etc. After 16 clock pulses, the first data bit appears on the Qh pin of the lower shift register. Thus, to output any data, you need 16 clock pulses.

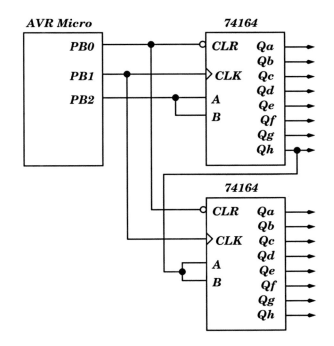

By changing the configuration of the connection between the two shift registers and the microcontroller, the number of clock signals can be reduced from 16 to 8, but at the cost of an extra microcontroller pin, as shown next. The data input to each of the shift registers is independently set by the microcontroller. The clock is generated commonly for the two shift registers by the microcontroller. This requires only eight clock pulses to output 16 bits of data on the two shift registers.

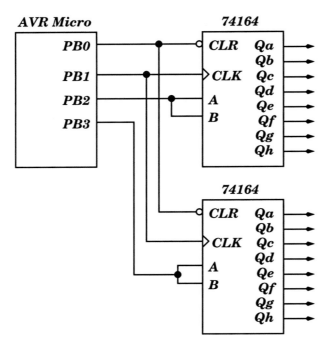

Compared to the configuration shown in the previous illustration, the scheme shown here requires eight clock signals, since the data input to the two shift registers is independently set by the microcontroller.

Let us apply the knowledge gained so far to implementing real circuits. One of the most common LED configurations is the 5 × 7 dot matrix display. The internal configuration of such a display is shown in the illustration at the top of the next column.

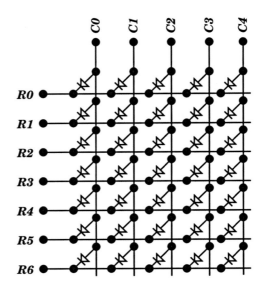

The LEDs' anodes are connected to the columns and the cathodes to the rows. However, 5 × 7 displays with anodes to rows and cathodes to columns are also available. One needs to be careful about the configuration when designing 5 × 7 LED matrices.

The next illustration shows the way the display matrix could be connected to an AVR microcontroller. The display has the anode connected to the columns and the cathode connected to the rows. In this design, a feature of

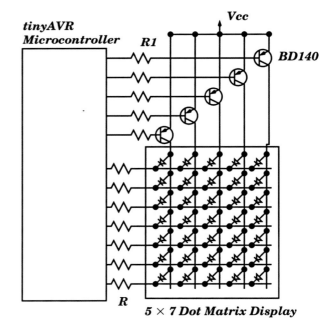

5 × 7 Dot Matrix Display

the AVR microcontroller, that every pin is capable of sourcing or sinking up to 40mA, is used. The anodes are connected to the Vcc, one at a time, through PNP transistors.

Depending upon which LEDs in a given column need to be turned on, the corresponding rows are connected to ground through the AVR microcontroller pins. Resistor "R" is set to limit the current through the LED to 40mA, since that is what each AVR microcontroller pin can tolerate. To turn a column on, the corresponding port pin connected to the base of the transistor is set to logic "0." This turns the PNP transistor on and allows the current to flow through the enabled LEDs. Since there are five columns, the duty cycle for the current through the LED in a given column is 20%. Thus, the average current through the LED is 20% of 40mA, that is, 8mA. Such a scheme is, therefore, suitable only for small displays. For

larger displays, which have higher average and peak current ratings, a different circuit would have to be used to facilitate the larger current.

The bottom image shows a scheme to increase the peak current through the LEDs by using NPN transistor switches. ULN 2003 is an IC with seven NPN drivers with logic input capability, and each output of the IC can handle up to 500mA of current. Now with the ULN driver in place, the value of current-limiting resistor R can be much smaller to allow larger peak current though each LED.

The next illustration shows a scheme similar to the one shown earlier (without ULN 2003), except this uses NPN transistors. In this scheme, each row is enabled (with a logic "1" at the input of the NPN transistor base) and, depending upon which LEDs in that row need to be turned on, the anodes are set to logic "1." The value of the current-

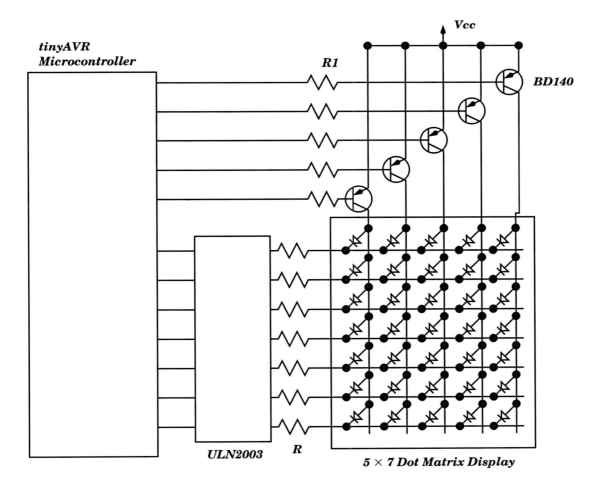

5 × 7 Dot Matrix Display

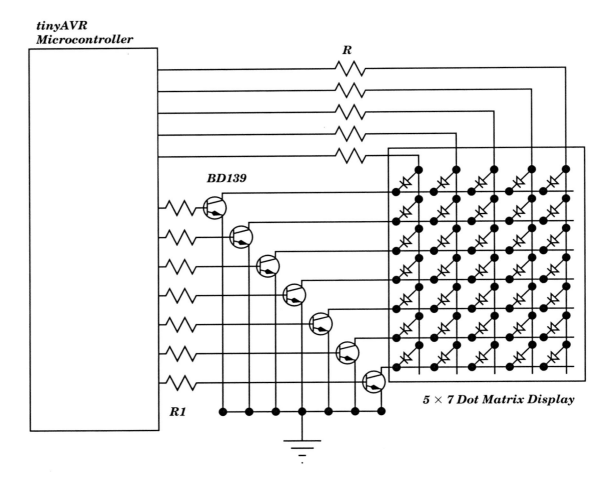

tinyAVR Microcontroller

R

BD139

R1

5 × 7 Dot Matrix Display

limiting resistor is still set to restrict the current to less than 40mA since the current is directly sourced by the microcontroller pins. However, in this scheme, the duty cycle is 1/7, since there are seven rows, and the average current through the LED is about 6mA—much lower than the average current through the LEDs in the scheme shown earlier with PNP transistors; however, this scheme uses NPN transistors and may be suitable for smaller size displays.

The image on the top of the next page shows a scheme using shift registers and a decoder to connect a 16 × 16 dot matrix LED display to an AVR microcontroller. The block diagram does not show the current boost transistors, which would be required on the high side as well as the low side of the display matrix.

The unique feature of this scheme is the use of a 4-to-16 decoder, 74154. The 74154 decoder IC has 16 active low outputs, which would be useful to enable PNP (or PMOS) transistors. An additional signal from the AVR microcontroller is used to disable the decoder using the G1/G2 enable signals. A pair of cascaded shift registers are used to provide 16 output signals to drive the 16 rows of the display. Enabling a particular column of the display is easily achieved by setting the inputs of the decoder to the required value; if the leftmost column (column number 0) needs to be enabled, then the decoder inputs ABCD = '0000'. For enabling column number 1, the ABCD input is set to "0001" and so on. To enable the decoder, G1/G2 input is set to logic "0." If all the columns need to be disabled, then G1/G2 input is set to logic "1." While changing the columns, the decoder is first

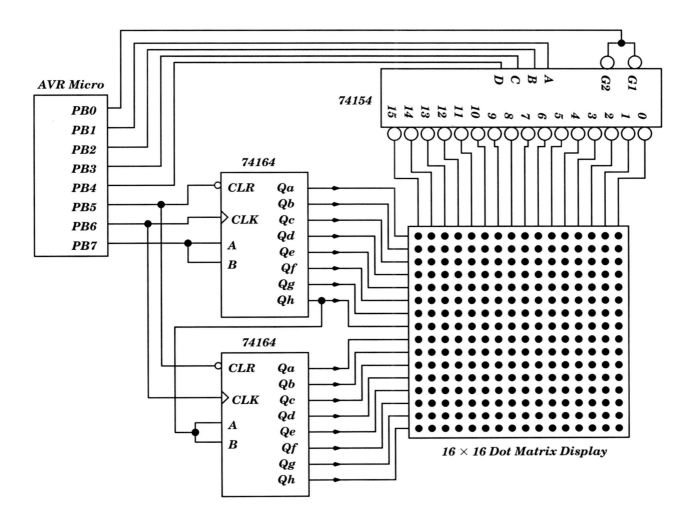

16 × 16 Dot Matrix Display

disabled by setting the G1/G2 input to logic "1" and then the new values for the 16 rows are shifted into the shift registers (remember that it will require 16 clock signals to output the 16 row values), and then the new column is enabled by appropriately setting the decoder inputs ABCD, and then the decoder is enabled by setting G1/G2 signal to logic "0." The block diagram shown in this illustration will also require 16 PNP transistors (or P-channel MOSFETs) at the output of the decoder and 16 NPN transistors (or N-channel MOSFETs) at the output of the shift registers.

Also, although the three illustrations prior to the one above show methods to control a 5 × 7 dot matrix display using an AVR microcontroller, these methods can just as well be applied to seven-segment and alphanumeric displays, which are also common. The next illustration shows the seven-segment display (left) and two types of alphanumeric displays (center and right). Each segment of the display is labeled with a letter. For a seven-segment display, the segments are labeled A through G and the decimal point is labeled dp. Actually, for a seven-segment display, there are eight segments, including the decimal point, but these displays are commonly referred as seven-segment displays in popular literature and datasheets. The alphanumeric displays are of two types with 14 segments and 16 segments, excluding the decimal point, as seen on the top of the next page.

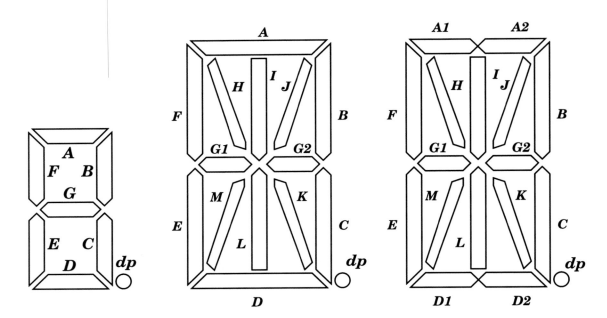

The next image shows the arrangement of the LEDs in a seven-segment display. Each display has a common signal, either the anode or the cathode. Thus, the seven-segment displays are available either in a common-anode or a common-cathode configuration. Similarly, the alphanumeric displays are also arranged in a common-anode or common-cathode configuration.

Each 5 × 7 dot matrix display can be replaced with five seven-segment displays. The following illustration shows the technique to control up to eight seven-segment displays. The scheme shown

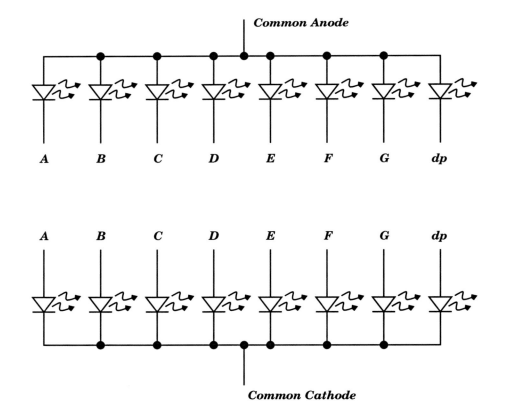

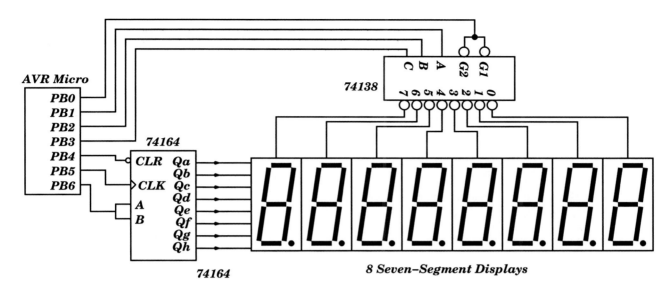

74164   **8 Seven–Segment Displays**

here uses a 74164 shift register to drive the segments of the seven-segment display and a 3-to-8 decoder 74138 to drive the common anodes of the seven-segment displays. Again, the current driver transistors have been excluded from the figure and, for a real-world example, would require appropriate PNP and NPN transistor drivers.

Charlieplexing

Using Charlieplexing as a method to multiplex LED displays has attracted a lot of attention recently, due to the fact that it allows one to control N*(N – 1) LEDs using N I/O lines. Compared to Charlieplexing, the standard multiplexing technique described in the previous sections controls a much smaller number of LEDs. Table 3-1 lists the number of LEDs that can be controlled by both the Charlieplexing method and the standard multiplexing method by splitting the available number of N I/O lines in suitable numbers of rows and columns. Table 3-1 also shows the duty cycle of the current that flows through the LEDs when it is turned on.

Clearly, Charlieplexing allows many more LEDs to be controlled with a given number of I/O lines.

TABLE 3-1	Comparison Between Charlieplexing and Multiplexing			
N (Number of I/O lines)	Maximum number of LEDs controlled with multiplexing	Duty cycle with multiplexing	LEDs controlled with Charlieplexing	Duty cycle with Charlieplexing
2	2	100%	2	50%
3	3	100%	6	16.67%
4	4	50%	12	8.33%
5	6	50%	20	5%
6	9	33%	30	3.33%
7	12	33%	42	2.4%
8	16	25%	56	1.78%
9	20	25%	72	1.38%
10	25	20%	90	1.11%

However, the downside to this technique is the reduced duty cycle of the current that flows through the LEDs; thus, to maintain a given brightness, the peak current through the LEDs must be increased proportionately, which can quickly reach the maximum peak current limit of the LED. Nonetheless, Charlieplexing is a feasible technique for up to ten I/O lines, allowing up to 90 LEDs to be controlled. Controlling an equivalent number of LEDs using the standard multiplexing technique would require 19 I/O lines.

Charlieplexing exploits a feature of programmable digital I/O lines available on modern microcontrollers, that is, the ability to hold an I/O line in high impedance state, commonly referred to as the "Z" state. To understand the operation of a Charlieplexed display, let us refer to the image here, which shows a microcontroller with just three I/O pins and six LEDs connected as shown.

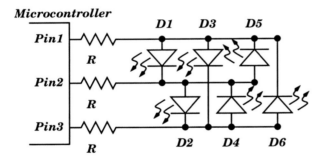

To turn LED D1 on, Pin1 is set to "1" and Pin2 is set to "0" while Pin3 is set to "Z," the tri-state of a logic output pin. Most modern microcontrollers such as the AVR allow each and every output pin of its ports to be operated in three states: logic "1," logic "0," and "Z." Similarly, to turn LED D2 on, Pin2 is set to logic "1," Pin3 is set to "0," and Pin1 is set to "Z." Table 3-2 shows the state of the pins to enable each of the LEDs. The way a Charlieplexed display is controlled is similar to the multiplexed display.

TABLE 3-2	Scheme of Charlieplexing		
LED Number	Pin1	Pin2	Pin3
D1	"1"	"0"	"Z"
D2	"Z"	"1"	"0"
D3	"1"	"Z"	"0"
D4	"Z"	"0"	"1"
D5	"0"	"1"	"Z"
D6	"0"	"Z"	"1"

However, unlike in the multiplexed display, where an entire row (or column) of LEDs is enabled, in Charlieplexed display, one LED at a time is enabled. Thus, the average current through the LEDs = I(peak)/X, where X is the total number of LEDs and X = N*(N − 1), where N is the number of pins. Also, it is difficult to include additional current-boosting switches in a Charlieplexed display, and the raw capabilities of the microcontroller determine the peak current. For AVR microcontrollers, the maximum current that a pin can source or sink is 40mA. In the illustration, two resistors of value R come in the path of the LED current. Assume a supply voltage Vcc and LED turn-on voltage V(led). Thus

$$I(max) = (Vcc − V(led))/2R$$

Since I(max) is 40mA, then R = (Vcc − V(led))/2*I(max). For Vcc = 5V and red LED (V(led) = 2V), the value of R = 37.5 Ohms, so a 39-Ohm standard resistance can be used for R.

Using Charlieplexing to control multiple LEDs is useful with small microcontrollers with a limited number of pins. However, since microcontrollers are not designed to supply large source/sink currents, it is not recommended to use this technique with more than six pins, that is, to control no more than 30 LEDs. With 30 LEDs, the average current would be limited to just over 1mA, which may be only suitable for small low-range applications. We use this technique in some of the projects in this chapter.

Project 6
Mood Lamp

The mood lamp is a project similar to the RGB LED color mixer project in the previous chapter (Project 3), in that the mood lamp also uses RGB LEDs. However, the purpose of the mood lamp is to create ambient light of any required color to help you meditate or relax, or simply to set the ambient light depending upon your mood. In the color mixer project, a single RGB LED was used, and the intensity of the individual red, green, and blue LEDs' intensity was set by corresponding potentiometers to create a custom color. Each potentiometer set the intensity to a value between 0% and 100% in 256 individual levels, so a total of 16 million colors could be generated. The mood lamp also uses RGB LEDs. However, instead of a single RGB LED, the mood lamp is capable of controlling a large number of RGB LEDs, since the objective is to provide ambient lighting. Also,

the mood lamp does not provide individual intensity control over the LEDs; instead, it allows you to select a particular color from a table of colors preset in the internal nonvolatile memory of the system. Each color entry in the nonvolatile memory, in turn, consists of three intensity values for the red, green, and blue LEDs. The intensity of the LED is controlled through PWM with a five-bit resolution; thus, for each color, the intensity from minimum to maximum has 32 levels. The bottom illustration shows the block diagram of the mood lamp.

The raw DC voltage should be 12V since this voltage is used to drive the LEDs, as explained later. This project uses commonly available RGB LED strings. A photograph of such an LED string is shown later in Figure 3-5. The spool consists of cascaded building blocks. There are about ten blocks in about three feet of LED string. Each building block consists of a set of red, green, and blue LEDs. Furthermore, each set consists of three LEDs in series and each series combination of LEDs has a current-limiting resistor, as shown on the next page.

Raw DC Voltage

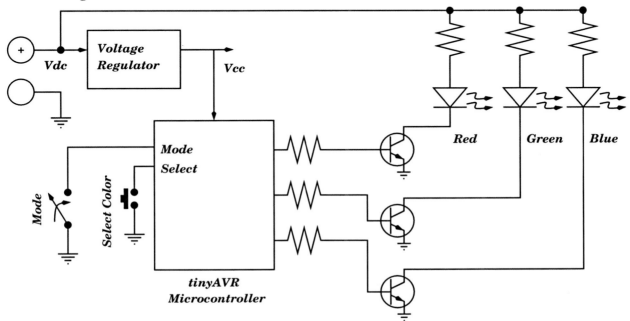

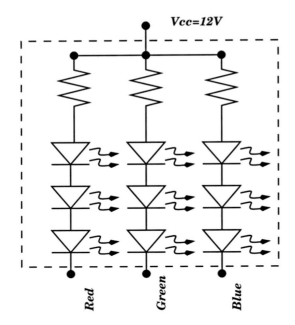

Vcc=12V

Red   Green   Blue

The system operates in two modes: continuous color change mode and fixed color mode, selected using the "mode" switch. In the fixed color mode, the actual color can be selected using the "select color" switch. The microcontroller is powered with a 5V regulator, which derives its voltage input from the raw DC voltage. Although the project uses commonly available LED strings, if required, individual high-power, high-brightness LEDs could also be used. A convenient configuration would be similar to the arrangement shown in the illustration, and would consist of three red, three green, and three blue LEDs, each of 1W power rating. For such a configuration, a series current-limiting resistor for red, green, and blue LEDs should be appropriately chosen to restrict the current to 300mA each.

Design Specifications

The aim of this project is to develop a user-selectable ambient colored lighting system using RGB LEDs. The controller should be capable of handling about 10W of power for each of the three colors. Also, the system should offer a mode to change the color gradually and to cycle through all the available colors. An external DC power supply of 12V is assumed to be available. The system should be able to handle commonly available LED strings or to use individual 1W high-brightness LEDs.

Design Description

Figure 3-2 shows the schematic diagram of the project. The voltage regulator used is again LM2940 with 5V output. So, input voltage can vary from around 6V to 20V. Diode D1 is a Schottky diode (IN5819) used as a protection diode, as explained earlier. Capacitors C5 and C7 are used to filter the spikes and unwanted noise in the power supply. C4 and C6 are used to stabilize the output of LM2940. C1 and C3 are soldered near the supply pins of the microcontroller to further decouple the noise arising in the circuit. LED1 is a 3-mm power on/off indicator red LED. The microcontroller used is the ATtiny861 microcontroller. It has three hardware PWM channels on Timer1 required for driving the three transistors T1, T2 and T3. The transistors (2SD789) used are NPN transistors with current sinking capability of 2 Amperes. They are connected in open-collector configuration. Toggle switch SW3 is used to select either of the two modes—continuous or discrete—and push button switch S1 is used to change the color in discrete mode. The 3-mm LED (2) is used to indicate the present selected mode. SL2 is the connector for mood-generating LEDs (either an RGB spool or high power LEDs). It is a four pin connector. The first pin is used to power the anodes of the LEDs and the other three pins are used to drive the cathodes of the Red, Blue, and Green LEDs. The current-limiting resistors should be externally connected. The voltage going to the LEDs is the unregulated VRAW so that large amount of current can be drawn from it. Otherwise, the driving capability would have been limited at the maximum output current capacity of the voltage regulator.

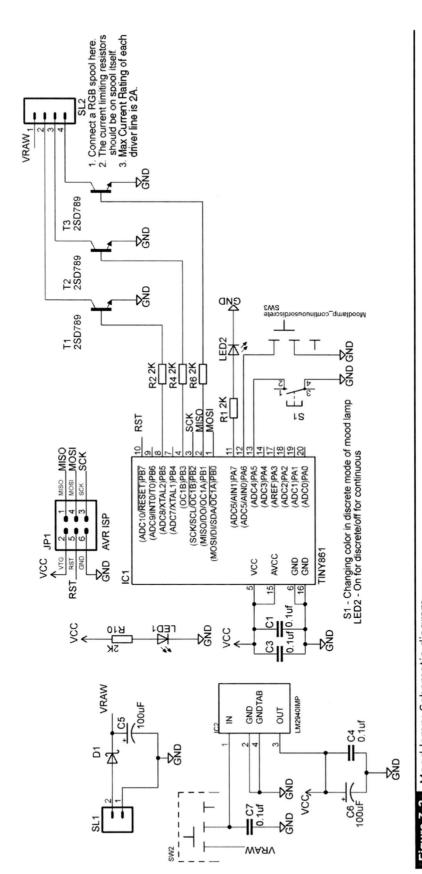

Figure 3-2 Mood lamp: Schematic diagram

The source code of the project reads the values of the switches, and if the mode is continuous, it continuously changes the duty cycle on three hardware PWM channels. In discrete mode, it waits for the press and release of S1 for each update of the PWM duty cycle values. The resolution of each PWM channel is set to be five bits.

Fabrication

The board layout in EAGLE, along with the schematic, can be downloaded from www.avrgenius.com/tinyavr1.

The board is routed in the solder layer with a few jumpers in the component layer. The component and solder sides of the soldered board are shown in Figure 3-3 and Figure 3-4,

respectively. We have used a 16-foot RGB spool wrapped around a glass tube as a mood generator lamp. It is shown in Figure 3-5. Figure 3-6 shows the RGB spool connected to the main board and displaying one of the combinations of colors.

Design Code

The compiled source code, along with the MAKEFILE, can be downloaded from www.avrgenius.com/tinyavr1.

The code runs at a clock frequency of 8 MHz. The controller is programmed using STK500 in ISP programming mode. The important sections of the code are shown on the opposite page.

Figure 3-3 Mood lamp: Component layout

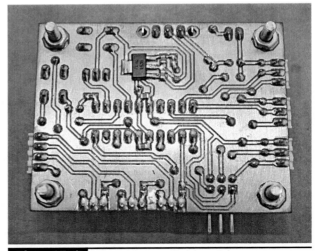

Figure 3-4 Mood lamp: Solder side

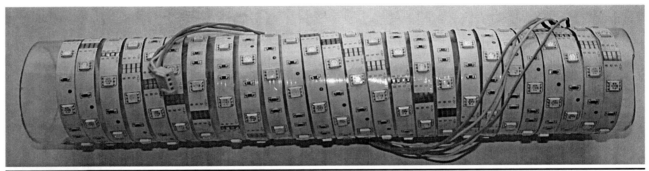

Figure 3-5 Mood lamp: RGB spool

```
    else if(mode==DISCRETE)
    {
      while((PINA&(1<<5))&&(mode==
      DISCRETE));
        //wait till switch is pressed
      _delay_ms(30);
      while((!(PINA&(1<<5)))&&(mode==
      DISCRETE));
        //wait till switch is released
      _delay_ms(30);
    }
  }
  }
 }
}
```

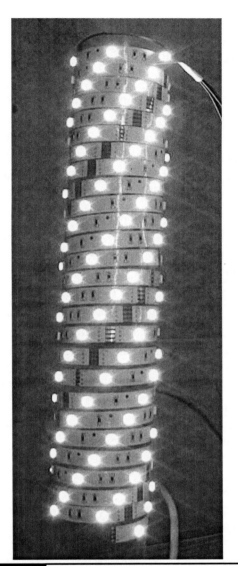

Figure 3-6 Working mood lamp with RGB spool

```
while(1)
{
 for(i=0;i<32;i++)
 {
  OCR1A = i;
  for(j=0;j<32;j++)
  {
   OCR1B = j;
   for(k=0;k<32;k++)
   {
    OCR1D = k;
    if(mode==CONTINUOUS)
    _delay_ms(500);
```

This is the main infinite loop of the program. It has three cascaded for loops inside it that change the intensity level on each hardware PWM channel through **OCR1A**, **OCR1B**, and **OCR1D** registers. The **mode** can either be DISCRETE or CONTINUOUS, and is changed by using the pin change interrupt on the pin connected to the toggle switch. If the mode is continuous, the loops run continuously, with a delay of 500 ms between successive iterations. In discrete mode, the code waits for the press and release of the switch on the fifth pin of PORTA through two while loops. As soon as the mode is changed from discrete to continuous, the code exits the while loops.

```
ISR(PCINT_vect)
{
 _delay_ms(30);//debounce
 GIFR = 1<<PCIF;
   //Clear flag set due to switch
   //bounce
 mode =
   mode==CONTINUOUS?DISCRETE:CONTINUOUS;
 if(mode==CONTINUOUS)
 PORTA&=~(1<<7);//LED off
 else
 PORTA|=(1<<7);//LED on
}
```

This part of the program is the interrupt service routine for the pin change interrupt, which is called each time the state of the toggle switch is changed. It changes the value of the mode, and the infinite loop discussed earlier changes its execution accordingly. CONTINUOUS and DISCRETE are macros declared at the start of the program. Depending on the mode, the state of the LED changes, as discussed earlier.

Apart from this, the rest of the code includes the initialization of Timer1 and its hardware PWM channels.

Working

By default the mode of the mood lamp is continuous. So it gradually changes the colors on the connected LEDs. If the intensity of the LEDs is large enough, you can feel the ambient environment around you changing. In discrete mode, you can use push button to set the color of your choice.

Project 7
VU Meter with 20 LEDs

A VU meter is often seen on audio equipment to indicate the "Volume Unit." Older audio equipment had analog-type indicators, while the modern ones

often have LED-based indicators. The purpose of the VU meter is to get a sense of the perceived loudness of the signal. Apart from audio equipment, a VU meter can be used in any application to measure intensity levels. LED-based VU meters are so common and popular that semiconductor manufacturers offer dedicated integrated circuits to measure external signals and display the output on external LEDs. The common arrangement of an LED-based VU meter is to arrange all the LEDs in a single column or row. One such popular IC is the LM3914 from National Semiconductors. The LM3914 is an interesting and popular dot/bar display driver used for sensing analog voltage levels and displaying them on ten LEDs, providing a linear analog display. The LM3914 has been around for more than 20 years. We wanted an output display solution with more than ten levels. It is possible to cascade multiple LM3914s for more than ten levels, but even so, the basic character of the device is only a linear display of the input voltage. Given this restriction, we had to find an alternate solution.

This illustration shows the operation of a VU meter in bar mode and dot mode. In the bar mode, with increasing signal level at the input, more and more LEDs light up from bottom to top. In the dot

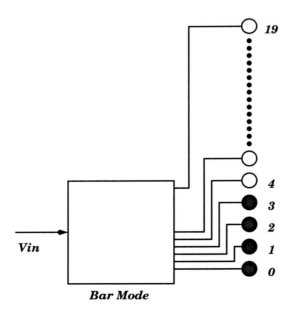

Bar Mode

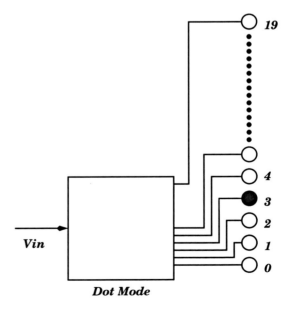

Dot Mode

mode, as the input signal increases, a single LED representing the higher signal lights up. The LEDs are arranged from bottom to top. The LED at the bottom indicates a lower signal level as compared to an LED at the top.

Design Specifications

The aim of this project was to create a versatile, 20-level VU meter using as small a microcontroller as possible. Interfacing 20 LEDs to a microcontroller is possible with just five I/O pins. Thus, an eight-pin microcontroller that offers up to six I/O pins would be a perfect component to implement this project. Also, given the flexibility offered by a program, the relationship between the input signal and output LEDs could be tailored to meet any specific requirement. For example, the illustration shown next shows a logarithmic scale. A linear scale could also be used by reprogramming the microcontroller with suitable code. To display the input voltage in a linear fashion, the input signal would be divided with a constant number. For the logarithmic scale, a lookup table could be easily employed.

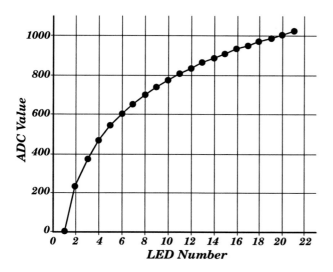

Design Description

Figures 3-7 and 3-8 show the schematic diagrams of the project. Diode D1, a Schottky diode (IN5819), is again used as a protection diode. Capacitor C1 is used to filter the spikes and unwanted noise in the power supply. C2 is a decoupling capacitor, explained earlier. The microcontroller used is the ATtiny45 microcontroller. There is no voltage regulator on this circuit, so input voltage can only vary from 4.5 to 5.5V. LEDs 1 through 20 are the 20 LEDs arranged in the form of a bar. Since 20 LEDs arranged in a bar at a suitable distance increase the size of the board beyond the limits permitted by the free version of EAGLE, the circuit has been split into two parts. The first part is shown in Figure 3-7. It has all the control circuitry and ten LEDs. The remaining ten LEDs are arranged separately, as shown in Figure 3-8. The PCBs of the two circuits have been designed separately, with a provision to connect the two such that they give the appearance of one bar of 20 LEDs.

Twenty LEDs have been Charlieplexed using five I/O pins of the microcontroller. SL2 is a three-pin connector for providing the input to the ADC channel of the controller. The input can be variable DC voltage provided by a potentiometer with its three terminals connected to VCC, GND, and PB5. It can also be an input waveform connected between PB5 and GND. If the input is a waveform, its minimum and maximum amplitude levels must be limited to 0(GND) and VCC, respectively.

The source code of the project takes a moving average of ten consecutive values of ADC readings, divides it into 21 levels (0–20), and switches on the equivalent number of LEDs.

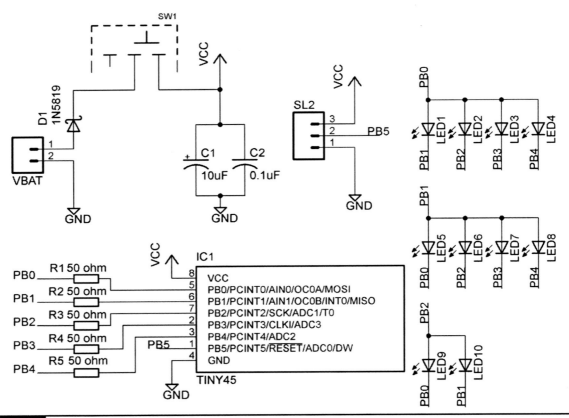

Figure 3-7   VU meter: Schematic diagram 1

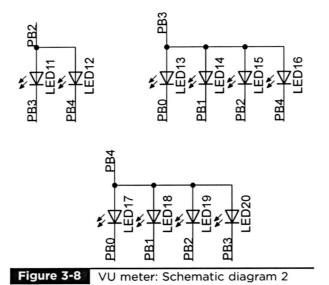

Figure 3-8   VU meter: Schematic diagram 2

Fabrication

The board layouts in EAGLE, along with the schematics, can be downloaded from www.avrgenius.com/tinyavr1.

Both the boards are routed in the solder layer with a few jumpers in the component layer. The two boards are joined by connecting the corresponding extended tracks on both the PCBs. A complete assembly of the board is shown in Figure 3-9.

Design Code

The compiled source code, along with the MAKEFILE, can be downloaded from www.avrgenius.com/tinyavr1.

The code runs at a clock frequency of 8 MHz. The reset pin (PB5) of the microcontroller is used

Figure 3-9   VU meter: Soldered board

as an ADC input pin. So the reset function on that pin needs to be disabled by programming the RSTDISBL fuse. Once that fuse is programmed, the controller can't be programmed using the ISP interface. Hence, high voltage serial programming (HVSP) mode has been used to program the controller using STK500. The important sections of the code are explained here:

```
while(1)
{
  //shift the values
  for(i=0;i<9;i++)
  {
    adcreading[i]=adcreading[i+1];
  }
  //take new reading
  adcreading[9] = read_adc();
  //Find the sum and perform
  //quantization
  adcsum = 0;
  for(i=0;i<10;i++)
  {
    adcsum = adcsum + adcreading[i];
  }
  //Divide sum of 10 ADC reading from 0
  //to 2550 into 21 levels(0 to 20)
  adcsum = adcsum/122;
  if(level>adcsum)
  {
    for(i=adcsum;i<level;i++)
    {
      statusonoff[i]=0;
    }
  }
  else if(level<adcsum)
  {
    for(i=level;i<adcsum;i++)
    {
      statusonoff[i]=1;
    }
  }
  level=adcsum;
}
```

This is the main infinite loop of the program. It deletes the earliest ADC reading from the buffer **adcreading**, shifts the remaining values, and takes a new ADC reading. The ADC of the ATtiny45 has been used in eight-bit resolution. Then it indirectly calculates the average by summing all the readings in the buffer and dividing the sum into 21 levels (equivalent to first dividing the sum by 10 and then dividing the resultant average into 21 levels). Then, the required numbers of LEDs are either switched on or off, depending upon the previous level (or previously on LEDs). Finally, **level** is updated with the present reading. The Charlieplexing code, as explained in the next section, has been written in such a way that if at any instant of the program **statusonoff[p]** is set to 1, the pth position LED starts to glow, and vice versa. The mapping of level to the number of LEDs glowing has been done in a manner that no LED is switched on for 0 level and all 20 LEDs are switched on for the highest (20) level.

```
//Overflow routine for timer0
ISR(TIM0_OVF_vect)
{
  DDRB=0;
    PORTB=statusonoff[count])
      <<pgm_read_byte(&anode[count])
      |0<<pgm_read_byte(&cathode[count]);
  DDRB = 1<<pgm_read_byte(&anode[count])
      |1<<pgm_read_byte(&cathode[count]);
  count++;
  if(count==20)
  count=0;
}
```

This part of the program is the interrupt service routine for Timer0 overflow. Thus, it is routinely called at a specific rate. It handles the Charlieplexing of 20 LEDs. **Count** is used to keep track of the LED that is operated upon in the present slot. The corresponding value stored in **statusonoff [count]** is given to its anode, and 0 is given to the cathode. Further, the anode and cathode pins are declared as outputs by updating the DDRB register. Thus, the LED is switched on if **statusonoff [count]** is 1, and vice versa. This cycle is repeated for every LED.

The rest of the code consists of initializations for ADC and Timer0.

Working

We tested the device by soldering a 10K potentiometer to SL2. The DC average value was varied and its effect was noted on the LED bar.

Project 8
Voltmeter

This project and the next two projects are built around a common hardware circuit consisting of a two-and-a-half digit, seven-segment display using a mere eight-pin microcontroller. Instruments often use LED (or LCD) based seven-segment displays. A common display configuration is three and a half digit, which consists of three complete digits and a leading digit that can display "1" if required. Thus, a three-and-a-half-digit display can show a value from 0 to 1999. If the three-and-a-half-digit display has a negative sign, it can show values from –1999 to 1999. Higher resolution displays would have four-and-a-half-digit resolution, which would have a range of 0 to 19,999 or –19,999 to 19,999, which is a ten-fold increase in range over the three-and-a-half-digit display. For many applications, a smaller two-and-a-half-digit display may also be suitable. Instead of ready-built seven-segment displays, one can also build a seven-segment display using individual LEDs. The advantage of making such a display is that one can choose any color for the LEDs and any size of the display. The following illustration shows how a seven-segment display can be made using individual LEDs. Each segment of the display, except the decimal point, is made with at least three LEDs in parallel. Recall from the previous chapter that driving LEDs in parallel without current-sharing resistors is not a good idea. However, we are going to work against that

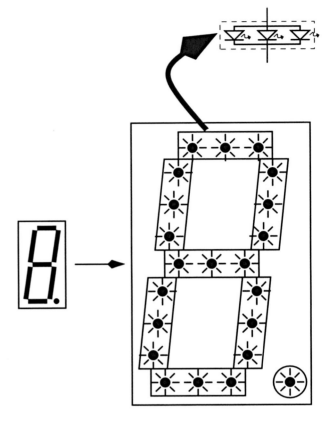

suggestion and go ahead and put three LEDs in parallel for each of the seven segments. It's a good idea to sort the LEDs based on intensity, although that is a manual process and can be time consuming.

Each seven-segment display has eight segments. Now, an eight-pin microcontroller has two power supply pins and six I/O pins. With five I/O pins, using Charlieplexing, one can control 20 LEDs. Thus, one can easily connect a custom-built two-and-a-half-digit display to an eight-pin microcontroller using just five I/O pins. The sixth pin can be used for other applications, for example, to read external analog voltage or digital input. The 20 LEDs that can be controlled using Charlieplexing are used to connect two seven-segment displays, which take up 16 LEDs. The remaining four LEDs are used to display the leading digit with a decimal point and the negative sign, as shown next.

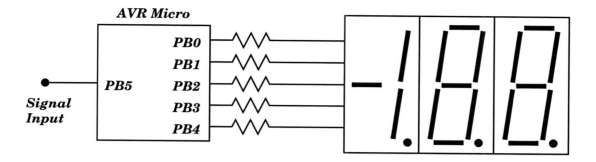

Design Specifications

The aim of the project is to design a single-range voltmeter using just an eight-pin microcontroller. The two-and-a-half-digit display is implemented using Charlieplexing, as explained in the previous section. The Tiny series of AVR microcontrollers has multiple channels on ADCs with ten-bit resolution. A ten-bit resolution means the ability to resolve 1 part in 1,024 parts. A two-and-a-half-digit display has a resolution of 1 part in 200 parts. So, the ten-bit ADC resolution is not a limitation for the voltmeter. For a single range, we chose 0 to 12V, which gives us a resolution of 0.1V on this display. The internal reference voltage of the microcontroller is 2.56V, so an external potential divider of 1:4.9 was chosen to give the range of approximately 12V. This can be suitably changed for any other range you desire.

Design Description

Figures 3-10 and 3-11 show the schematic diagrams of the project. Projects 8, 9, and 10 use the same circuit. C1 is a decoupling capacitor to remove the noise arising in the circuit, and is soldered near the supply pins of the controller. The microcontroller used is again the ATtiny45 microcontroller. There is no voltage regulator on this circuit, so the input voltage can only vary from 4.5 to 5.5V. Also, there is no capacitor to filter the spikes in the power supply. Hence, use of batteries due to their stable output is recommended. Twenty LEDs have been arranged in the form of two full

seven-segment displays (each requiring seven LEDs for digits and one for the decimal point), one half-digit display (which uses two LEDs to show either 0 or 1 and one for the decimal point), and one horizontal bar requiring one LED, which can be used to denote a minus sign or some other indicator. Since the digit segments of a seven-segment display are longer than the diameter of each 3-mm LED, three LEDs have been used in parallel to denote a single segment. This may lead to unequal current flowing in the segment bars and pointer LED. An additional resistor has been used in series with the decimal-point LEDs to mitigate the effect to some extent.

Twenty LEDs have been Charlieplexed using five I/O pins of the microcontroller. SL1 is a three-pin connector used for providing the input to the ADC channel of the controller. This input is different for Projects 8, 9, and 10. The second part of the circuit shown in Figure 3-10 is used for providing different inputs to the controller by connecting the input connector SL1 with any one of VOLT, LM35, and SIGNAL connectors. For this project, input comes by providing a voltage to be measured on SL2. This is then stepped down in the ratio of 4.9:1 and given to the ADC input channel of the microcontroller through the VOLT connector.

The source code of the project notes the ADC reading from the input channel, converts it into actual voltage, performs the rounding, and displays the result on seven-segment displays.

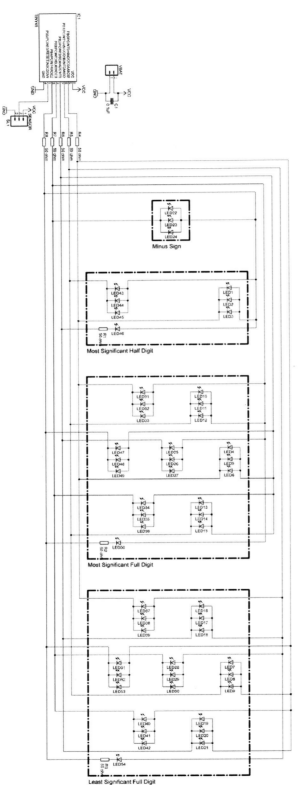

Figure 3-10 Autoranging voltmeter, Celsius and Fahrenheit thermometer, autoranging frequency counter: Schematic diagram 1

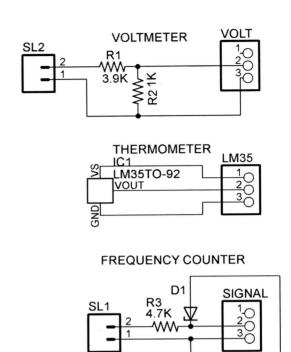

Figure 3-11 Autoranging voltmeter, Celsius and Fahrenheit thermometer, autoranging frequency counter: Schematic diagram 2

Fabrication

The board layouts in EAGLE, along with the schematics, can be downloaded from www.avrgenius.com/tinyavr1.

Both of the boards are routed primarily in the solder layer. The jumpers in the main board are not straight, so it is better to fabricate it on a double-sided board. There is no jumper in the second board, which is used to provide the inputs. Figures 3-12 and 3-13 show the top and bottom sides of main board. Figure 3-14 shows the assembled board for providing inputs.

Figure 3-12 Main board: Component side

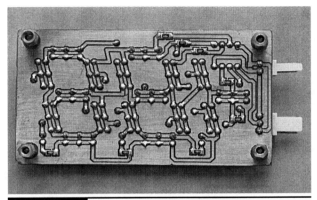

Figure 3-13 Main board: Solder side

Figure 3-14 Assembled board for providing inputs

Design Code

The compiled source code, along with the MAKEFILE, can be downloaded from www.avrgenius.com/tinyavr1.

The code runs at a clock frequency of 8 MHz. In this case, the reset pin (PB5) of the microcontroller is used as an ADC input pin. So, the reset function on that pin needs to be disabled by programming the RSTDISBL fuse. Hence, HVSP mode has been used to program the controller using STK500. The important sections of the code are explained here:

```
while(1)
  voltage = read_adc();
  //Here the voltage is 100 times the
  //actual voltage
  //Perform the Autoranging and scale
  //the result
  if(voltage<=199)
  {
   point = 1;
  }
  else if(voltage<1995)
  {
   if(voltage%10>=5)
   {
    voltage = voltage +10;
   }
   voltage = voltage/10;
   point = 2;
  }
  else
  {
   if(voltage%100>=50)
   {
    voltage = voltage+100;
   }
   voltage = voltage/100;
   point = 3;
  }
  c=voltage/100;
  voltage = voltage%100;
  d= voltage/10;
  voltage = voltage%10;
  e = voltage;
  display(c,d,e,point,0);
  _delay_ms(100);
}
```

This is the main infinite loop of the program. It first reads the ADC reading using the **read_adc** function, which provides a value that is 100 times the input voltage. The ADC has been used in ten-bit resolution and the reference voltage used is the internal 2.56V bandgap reference to improve accuracy. Since there is a potential divider of 4.9:1 before the input is applied to the controller, the input voltage to be measured can vary from 0 to 2.56*4.9 equal to 12.544(~12)V.

Once the ADC reading has been taken, rounding is performed by first selecting the suitable location for the decimal point and then rounding off the result to three digits. Then, the digits are extracted and passed to the display function, which maps it to the statusonoff array. The role of this array is same as that in Project 7.

Working

Input voltage is provided to the VOLT connector, and its value can be seen on the seven-segment displays.

Project 9
Celsius and Fahrenheit Thermometer

This project uses the same hardware as in the previous voltmeter project, that is, the two-and-a-half-digit seven-segment display. Instead of the external potential divider circuit, this circuit uses a temperature sensor to convert temperature into voltage that is measured by the built-in ADC of the microcontroller and converted into the Celsius or Fahrenheit scale temperature reading, which is then displayed on the two-and-a-half-digit display. There are many types of temperature sensors. Common ones are the thermistor, thermocouple, and the silicon bandgap temperature sensor. A thermistor, which is a temperature-dependent resistor, is easy to use and is cheap. A simple

circuit, as shown next, can be used to convert the temperature value into voltage.

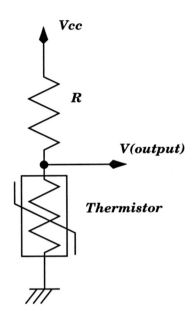

However, the variation in the resistance of the thermistor as a function of temperature is not linear. For accurate temperature measurement, a complex mathematical equation called the Steinhart-Hart equation needs to be applied. A thermocouple, on the other hand, is a useful temperature sensor quite suitable for large temperature measurements (such as those falling in the range of hundreds of degrees Celsius). A thermocouple, like the thermistor, is a nonlinear temperature sensor. It provides voltage output as a function of the temperature. A thermocouple also requires a polynomial approximation to convert the voltage output into the corresponding temperature reading. A silicon bandgap temperature sensor is the simplest to use. It provides direct and proportional output voltage (or current) corresponding to the temperature. In this project, we use such a bandgap sensor, the LM35 from National Semiconductors, which is also compatible with the TMP36 sensor from Analog Devices. The LM35 sensor provides 10mV per degree Celsius. Since the microcontroller has a ten-bit ADC with a range of 2.56V, it offers good resolution to measure the temperature with an accuracy of a fraction of a Celsius.

Design Specifications

The aim of this project is to measure voltage from a bandgap temperature sensor and to display temperature in Celsius and Fahrenheit, alternatively, on the two-and-a-half-digit seven-segment display. The power supply for the project is not included in the schematic diagram, and you can choose any suitable regulated power supply using a linear voltage regulator or even a DC-DC converter. The recommended voltage output of the regulator is 5V. Alternatively, four alkaline cells of 1.5V or even 4 NiMH cells of 1.2V, arranged in series, could be used.

Design Description

This project uses the same circuitry as in Project 8, but input is provided from the LM35 temperature sensor. It gives 10mV as output for one degree Celsius. This is provided to the ADC input channel and then displayed on the seven segments after autoranging.

The source, apart from displaying the temperature, also performs the conversion from Celsius to Fahrenheit. The mode for displaying the temperature in Celsius or Fahrenheit is toggled every 5 seconds. The horizontal bar is switched on for Fahrenheit mode and switched off for Celsius mode.

Design Code

The compiled source code, along with the MAKEFILE, can be downloaded from www.avrgenius.com/tinyavr1.

The code runs at a clock frequency of 8 MHz. Again, the reset function on PB5 needs to be disabled by programming the RSTDISBL fuse.

The important sections of the code are explained here:

```
While(1)
{
  temperature = read_adc();
  //Here the temperature is 100 times
  //the actual temperature
  if(temperature<=199)
  {
   point = 1;
  }
  else if(temperature<1995)
  {
   if(temperature%10>=5)
   {
    temperature = temperature +10;
   }
   temperature = temperature/10;
   point = 2;
  }
  else
  {
   if(temperature%100>=50)
   {
    temperature = temperature+100;
   }
   temperature = temperature/100;
   point = 3;
  }
  c=temperature/100;
  temperature = temperature%100;
  d= temperature/10;
  temperature = temperature%10;
  e = temperature;
  display(c,d,e,point,mode==FAH);
  _delay_ms(100);
}
```

This is the main infinite loop of the program. It first reads the ADC reading using the **read_adc** function, which provides a value that is 100 times the actual temperature. The temperature is either in Celsius or Fahrenheit, depending upon the mode. ADC has been used in ten-bit resolution, and the reference voltage used is the internal 1.1V bandgap reference to improve the resolution.

Once the ADC reading has been taken, calculation is performed by first selecting the suitable location for the decimal point and then rounding off the result to three digits. Then, the digits are extracted and passed to the display function, which maps it to the statusonoff array. The role of this array is the same as that in Projects 7 and 8. Timer0, apart from handling the Charlieplexing, also counts five seconds before toggling the mode between Celsius and Fahrenheit.

Working

The project continuously displays the temperature with the mode toggled every five seconds. A horizontal bar is used to denote the mode, as explained earlier.

Project 10
Autoranging Frequency Counter

A frequency counter is an instrument that measures the frequency of an external signal. The external signal could be analog or digital in nature. The bottom illustration shows the block diagram of a frequency counter.

The input signal is shown as an analog signal. The input amplifier and wave shaper amplify the signal and threshold it to convert it into a corresponding digital signal. The system has an accurate internal time base generator that enables a gate to pass the input signals to a counter chain. The duration of the time for which the gate is enabled depends upon the required resolution of the measurement. For example, to resolve the input signal to 1 Hz, the gate signal must be one second in duration. To resolve the signal frequency to 0.1 Hz, the gate signal must be ten seconds in duration. The output of the counter is then suitably displayed on an LED or LCD display.

If the frequency of the incoming signal is rather low, then one requires a gate signal of inordinately long duration to measure the frequency. One way to alleviate this problem is to use the period of the input signal itself as a gate signal and instead measure a high frequency generated by the internal time base generator, as shown in the illustration on the top of the next page. However, this method measures the period of the input signal, and thus the system shown in that illustration is a period counter. To convert the period into frequency, the system must perform a simple mathematical calculation.

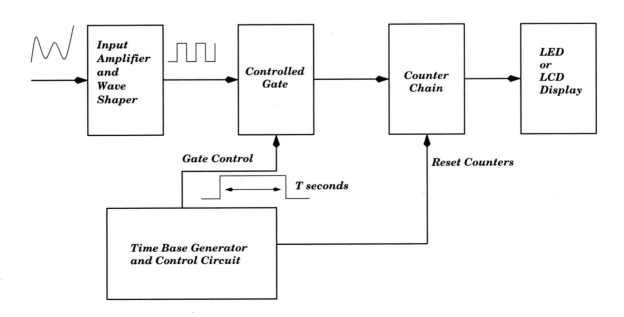

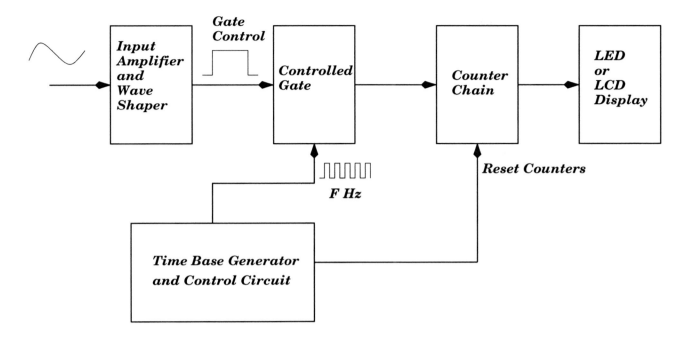

The bottom illustration shows the timing diagram of signals in a frequency counter. A period counter would have similar signals.

Design Specifications

In this project, the frequency of the input signal has to be measured and displayed on the two-and-a-half-digit seven-segment display. All the blocks shown in the first image for Project 10 are internally available in a microcontroller, and it can

also perform a mathematical calculation to convert from period to frequency as required. An autoranging frequency counter is an instrument that can automatically sense the frequency of the input signal and choose the optimum range. Our frequency counter has a display that has a range from 0 to 199. Thus, it can be used to display frequencies in the ranges shown in Table 3-3.

The decimal point is used to indicate which measurement range is being used. For the first range, the rightmost decimal point is switched on;

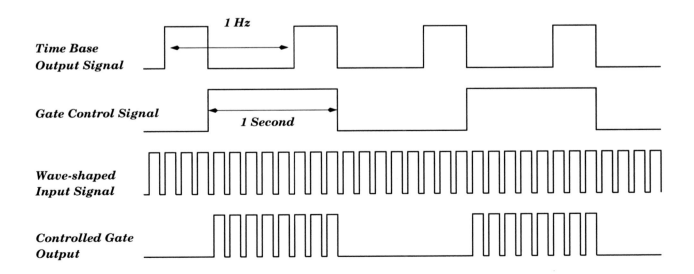

TABLE 3-3	Different Ranges of the Frequency Counter	
S. No.	Range	Gate signal period
1	0 to 199 Hz	1 second
2	0 to 1.99 KHz	0.1 second
3	0 to 19.9 KHz	0.01 second

for the second, the middle one; and for the third, the leftmost decimal point is switched on.

Design Description

This project uses the same circuit as that of Projects 8 and 9, but the input is provided from the SL1 connector of the circuit shown in Figure 3-10. D1 is a 5V Zener diode to limit the amplitude of the signal, and R3 is a current-limiting resistor. The input wave must be DC clamped. Its minimum amplitude should not be below ground level.

Design Code

The compiled source code, along with the MAKEFILE, can be downloaded from www.avrgenius.com/tinyavr1.

The code runs at a clock frequency of 8 MHz. The important sections of the code are explained here:

```
ISR(PCINT0_vect)
{
  edgecounter++;
}
```

This code excerpt is the interrupt routine for the pin change interrupt on PB5. So, **edgecounter** denotes double the frequency at the end of the gating period. The gating period has been kept to be one second throughout the code. At the end of each gating period, pulses per second are counted by dividing **edgecounter** by 2. The value of pulses per second is then rounded as in the previous two projects for display on the LED digits.

Working

The frequency is measured by applying a waveform on the SL1 connector. It may be noted that the frequency updating takes place after one second only.

Project 11
Geek Clock

Clocks of various shapes and sizes are commonly available. Our clock is based on LEDs and uses different types of LEDs to indicate time. The illustration on the following page indicates the block diagram of the clock.

This clock can indicate time to a resolution of one second. Three blocks of LEDs indicating hours, minutes, and seconds are used. Each block has a different number of LEDs based on the requirement. The LEDs representing the hours has two columns: higher-digit hours and lower-digit hours. Our geek clock displays time in a 24-hour format, so the hours will range from 00 to 23. Thus, the higher-digit hours will have no, one, or two LEDs lit. The lower-digit hours will have zero to nine LEDs lit at any given time. Similarly, the minutes are displayed in two blocks: higher-digit minutes and lower-digit minutes. The minutes range is 00 to 59. Six LEDs represent the seconds in a binary fashion. The reason why we call this a geek clock is that the LEDs representing the hours and minutes change randomly every five seconds. For example, if the time is 15 hours and 35 minutes, then any one LED in the higher-digit hours section will be lit. Any five LEDs out of nine in the lower-digit hours section will be lit. After five seconds, some other five out of the same nine LEDs would be lit. Thus, one would observe a changing LED lighting pattern. However, anyone trained to read the time in this fashion will be able to tell the time correctly.

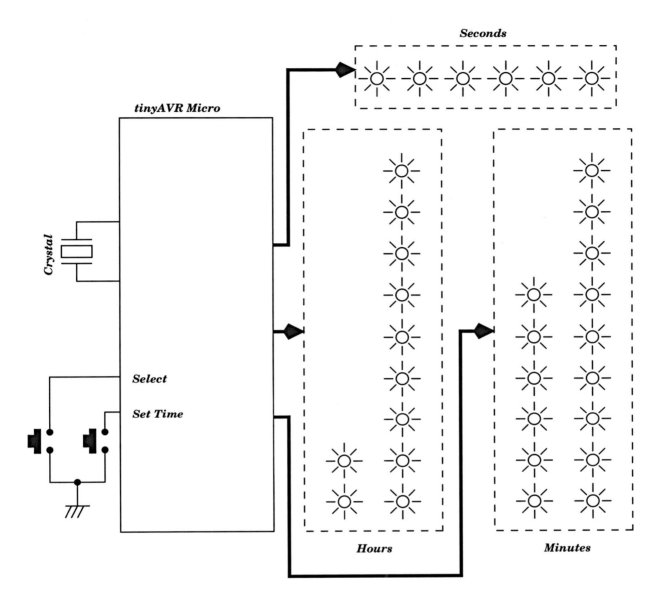

The illustration on the following page shows the time 15:35 displayed by the geek clock in two different ways.

Design Specifications

The aim of this project was to design a crazy geeky clock that should require special training to be able to tell the time. The time is displayed using different colored LEDs to indicate the hours, minutes, and seconds. To keep the time accurately, it was decided to use an external quartz crystal to provide the clock signal for the microcontroller.

Also, to interface so many LEDs, it was decided to Charlieplex the LEDs. However, rather than Charlieplex all the LEDs, the available I/O pins were isolated into three blocks: three pins to control the 6 LEDs for the seconds, four pins to control the 11 LEDs for the hours, and five pins to Charlieplex the 14 LEDs for the minutes. Also, two switches are provided to set the time. The power supply for the circuit is an on-board LM2940 linear low-drop-out voltage regulator. External batteries are used to power the circuit.

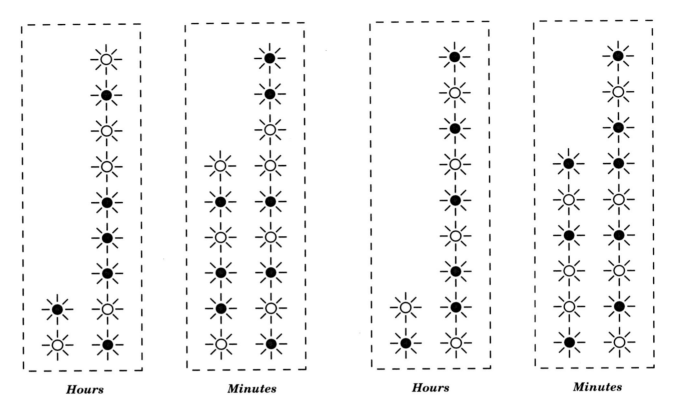

Hours   *Minutes*   *Hours*   *Minutes*

Design Description

Figure 3-15 shows the schematic diagram of the project. The voltage regulator used is again an LM2940 with 5V output, and so, input voltage can vary from around 6V to 20V. Diode D1 is a Schottky diode (IN5819) used as a protection diode, as explained earlier. C4 is used to filter the spikes and unwanted noise in the power supply, and C3 is used to stabilize the output of the LM2940. C1 and C2 are soldered near the supply pins of the microcontroller to further decouple the noise arising in the circuit. The microcontroller used is the ATtiny261 microcontroller. The LEDs have been Charlieplexed in three different groups: HOURS, MINUTES, and SECONDS. The SECONDS group has six LEDs and displays the seconds in binary format. LED6 is treated as the least significant bit and LED1 as the most significant bit. The MINUTES group has 14 LEDs. LED7 to LED11 represent the tens digit of minutes (0–5) when represented in decimal. LED12 to LED20 represent the ones digit of minutes.

Similarly, the HOURS group has 11 LEDs. LED21 and LED22 represent the tens digit, and LED23 to LED31 represent the tens digit of hours. It is recommended to solder different color LEDs for representing the seconds, minutes' tens digit, minutes' ones digit, hours' tens digit, and hours' ones digit.

R1 to R12 are the current-limiting resistors for the Charlieplexed LEDs and are of 50 Ohms each. R13 is used as a pull-up resistor to keep the RESET pin of the microcontroller pulled to Vcc. Q1 is a 1.8432-MHz crystal used to clock the microcontroller. In this project, the microcontroller is meant to keep time, which calls for a highly accurate oscillator. The default RC oscillator inside this microcontroller is not accurate enough to measure time. Therefore, a crystal is required.

Two switches to update the time have been connected in different configuration as compared to previous projects. This is due to the fact that the ATtiny261 has 15 I/O pins, excluding the reset pin. Out of these, 12 I/O pins are used for the LEDs and

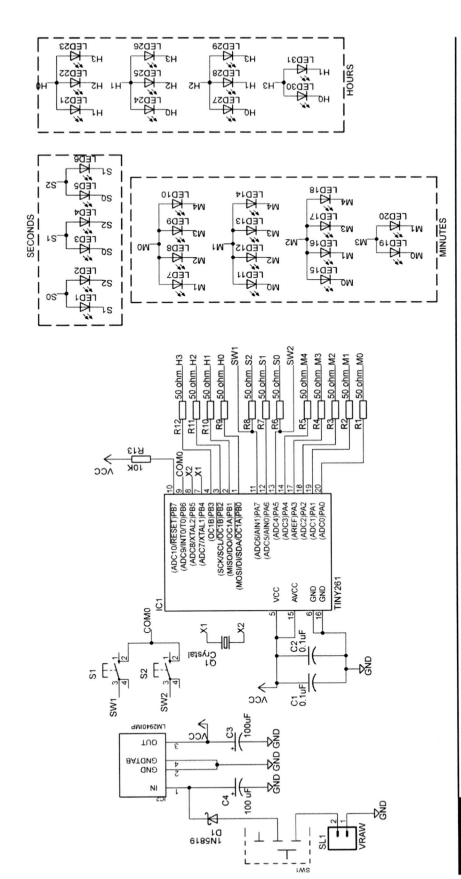

Figure 3-15 Geek clock: Schematic diagram

two for the crystal. This leaves us with one I/O pin and two switches. However, we know that I/O pins controlling the seconds LEDs have a fixed periodic pattern, that is, they would be either 0 or 1 after a fixed amount of time. Moreover, in Charlieplexing, only one LED is switched on at any instant, as explained before. We want one of the switches to be read by an I/O (COM0) pin of the microcontroller and the other to the ground. So, we have connected the other ends of the switches to the I/O pins controlling the seconds LEDs. When SW1 has logic 0, a change in logic level of COM0 is considered as a press of switch S1, and when SW2 has logic 0, a change in logic level of COM0 is considered as a press of switch S2.

The source code keeps accurate time by using timers. Also, switches S1 and S2 are used to update the time.

Fabrication

The board layout in EAGLE, along with the schematic, can be downloaded from www.avrgenius.com/tinyavr1.

The board is routed in the solder layer with a few jumpers in the component layer. A complete assembly of the board is shown in Figure 3-16.

Figure 3-16   Geek clock: Soldered board

Design Code

The compiled source code, along with the MAKEFILE, can be downloaded from www.avrgenius.com/tinyavr1.

The code runs at a clock frequency of 1.8432 MHz, which is obtained from an external crystal, so fuse bits have to be configured. The important sections of the code are explained here:

```
if((PORTA&0x80)==0)&&((DDRA&0x80)==0x80))
    //If PA7 declared as output and
    //equal to zero
{
 if(!(PINB&0x40))
 {
  switch_1_pressed = 1;
 }
 if(((PINB&0x40)==0x40)&&(switch_1_
    pressed==1))
 {
  switch_1_released = 1;
 }
}

if(!(PORTA&0x20)&&(DDRA&0x20)==0x20)
    //If PA5 declared as output and
    //equal to zero
{
 if(!(PINB&0x40))
 {
  switch_2_pressed = 1;
 }
 if(((PINB&0x40)==0x40)&&(switch_2_
    pressed==1))
 {
  switch_2_released = 1;
 }
}
```

This section of the code is one part of the interrupt overflow routine for Timer0 and is used to read the switches. For switch 1, it first checks whether PA7 has been pulled to 0 or not. If yes, then PB6 is read to detect the switch press or release. If PB6 is found to be 0, **switch_1_pressed** is given the value 1 to indicate the status of switch press. If PB6 is found to be 1 and **switch_1_pressed** is also 1, it means that the press and release of switch 1 has taken place and **switch_1_released** is updated to 1. The main code reads these and executes the corresponding function. A similar process happens for switch 2 when PA5 is pulled to 0.

```
void display(void)
{
 display_on = 0;
 u08 hour_ten,hour_one,min_ten,min_one;
 hour_one = hour%10;
 hour_ten = hour/10;
 min_one = min%10;
 min_ten = min/10;
 //display seconds
 for(u08 h = 0;h<=5;h++)
 {
  secondled[5-h] = ((sec&(1<<h)) ==
   (1<<h));
 }
 if(display_on1==1)
   //time to change random pattern
 {
 display_on1=0;
 //All hour and min leds off
 for(u08 o = 0;o<9;o++)
  {
   minuteled[o+5]=0;
    //clear minutes tens digit
   hourled[o+2] = 0;
    //clear hour tens digit
   if(o<5)
   minuteled[o] = 0;
    //clear minutes ones digit
   if(o<2)
   hourled[o] = 0;
   //clear hour ones digit
  }

 //display hour tens digit
 for(u08 o = 0;o<hour_ten;o++)
  {
   //generate random number from 0 to 1
   random = TCNT0;
   random = random%2;
   while(hourled[random] == 1)
    {
     random++;
```

```
     if(random==2)
     random = 0;
    }
   hourled[random] = 1;
  }

 //display hour ones digit
 for (u08 o = 0;o<hour_one;o++)
  {
   //generate random number from 2 to 10
   random = TCNT0;
   random = random%9+2;
   while(hourled[random] == 1)
    {
     random++;
     if(random == 11)
     random = 2;
    }
   hourled[random] = 1;
  }

 //display min tens digit
 for(u08 o = 0;o<min_ten;o++)
  {
   //generate random number from 0 to 4
   random = TCNT0;
   random = random%5;
   while(minuteled[random] == 1)
    {
     random++;
     if(random == 5)
     random = 0;
    }
   minuteled[random] = 1;
  }

 //display min ones digit
 for(u08 o = 0;o<min_one;o++)
  {
   //generate random numbers from 5 to 13
   random = TCNT0;
   random = random%9+5;
   while(minuteled[random] == 1)
    {
     random++;
     if(random == 14)
     random = 5;
    }
   minuteled[random] = 1;
  }
 }
}
```

The **display** function is the most critical component of the whole source code. It is called every second to update the seconds LEDs. However, if **display_on1** is equal to 1, it means that the present value of seconds is a multiple of five and that a new random pattern has to be generated on the minutes and hours LEDs. It uses timer0's value to calculate as many random numbers for the different group of LEDs as required. The code has been written in such a way that there cannot be any repetition of a random number generated within any group.

The rest of the code maintains the time and responds to user input through switches.

Working

There is no provision for keeping backup time in this project, so each time you power it on, it starts from 00:00 hours. You can increase minutes and hours by pressing and releasing S1 and S2, respectively. Please note that the switches don't respond until the I/O pins connected to these have logic 1. Time is displayed in 24-hour format.

Project 12
RGB Dice

If you play board games, you have to use dice. This project shows how to build an electronic die that not only produces a random number every time you press a switch, but also produces that number in a randomly chosen color. This is achieved by using integrated RGB LEDs to produce different color combinations. However, unlike the RGB color mixer or the mood lamp project, where PWM is used to modulate the intensity of individual red, green, and blue LEDs to generate a large number of color combinations, in this project, only three primary colors (red, green, and blue) and three secondary colors (yellow, orange, and purple) are used. The block diagram of the project is shown below.

The LEDs are arranged in the traditional dot pattern on a die. A switch is pressed and released to generate a random number by lighting up the LED pattern. Together with the lighting pattern to indicate the number, the color of the lights is randomly chosen from the six available colors,

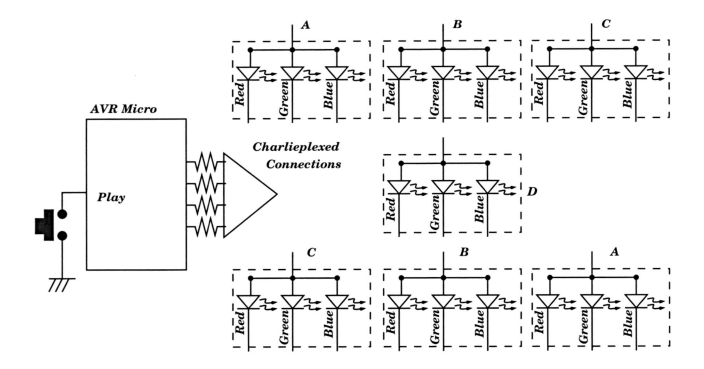

making it a unique and pleasing die to use in all your board games.

Design Specifications

The aim of the project was to design an electronic LED die. The LEDs were used to replace the traditional dots on a die. Each time a switch is pressed, the LEDs would light up to indicate the number. In addition, the color of the LED itself would be random, adding to the aura of the die. The die is powered by an external regulated power supply, although four alkaline cells of 1.5V in series or four NiMH cells of 1.2V in series could also be used.

Design Description

Figure 3-17 shows the schematic diagram of the project. The voltage regulator used is an LM3840 with 5V output, so the input voltage can vary from around 6V to 20V. Diode D1 is a Schottky diode (IN5819) used as a protection diode, as explained earlier. Capacitor C1 is used to filter the spikes and unwanted noise in the power supply. C2 is used to stabilize the output of LM3840. The microcontroller used is the ATtiny13 microcontroller. There are four groups of RGB LEDs arranged in a Charlieplexed fashion. There are 12 LEDs in all, controlled by four I/O pins. Each group contains two LEDs connected in parallel that are switched on simultaneously for all the patterns that may appear on a die, as explained before. However, a die is represented by seven dots. This requires three groups and one single LED, but in order to keep the current in all the LEDs constant, a dummy LED has been connected in parallel with the center dot. It is placed away from the display area and covered to hide its light. Switch S1 is used to change the number and colors of the LEDs. Pressing switch S1 changes the number, and releasing of the switch changes the color. R1 to R4 are current-limiting resistors.

The source code keeps running a counter, and whenever the switch is pressed, it captures the count value to generate a new number, and whenever it is released, it again captures the count value to generate a new color. There are six colors in total. Three are generated by switching on the red, green, or blue component of the required number of LEDs. The other three are generated by switching on any two components of the required LEDs. There is no PWM used in this code.

Fabrication

The board layout in EAGLE, along with the schematic, can be downloaded from www.avrgenius.com/tinyavr1.

The board is routed in the solder layer with a few jumpers in the component layer. The component and solder sides of the soldered board are shown in Figure 3-18 and Figure 3-19, respectively.

Design Code

The compiled source code, along with the MAKEFILE, can be downloaded from www.avrgenius.com/tinyavr1.

The code runs at a clock frequency of 9.6 MHz. The controller is programmed using STK500 in ISP programming mode. The important sections of the code are explained here:

```
//subroutine for switch pressed
void switchpressed(void)
{
 unsigned char b = 1;
 while((PINB&(1<<0)))
   //wait for switch to get pressed
 {
  if(b==6)
  b = 1;
  else
  b = b+1;
 }
 _delay_ms(20); //to prevent bounce
```

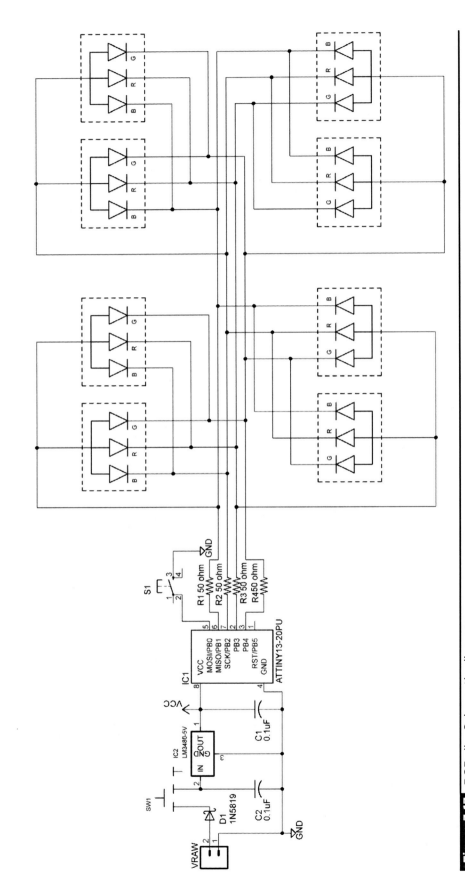

Figure 3-17 RGB die: Schematic diagram

Figure 3-18 RGB die: Component layout

Figure 3-19 RGB die: Solder side

```
statusonoff[1]=
    (b==4)||(b==5)||(b==6);
statusonoff[2]=
    (b==4)||(b==5)||(b==6)||(b==3);
statusonoff[3]= (b==2)||(b==6);
statusonoff[4]=
    (b==1)||(b==3)||(b==5);
}

//subroutine for switch released
void switchreleased(void)
{
unsigned char b = 1;
while(!(PINB&(1<<0)))
    //wait for switch to get released
    {
    if(b==6)
    b = 1;
    else
    b = b+1;
    }
    _delay_ms(20); //to prevent bounce
    coloronoff[0]=(b==2)||(b==4)||(b==6);
    coloronoff[1]=(b==3)||(b==4)||(b==5);
    coloronoff[2]=(b==1)||(b==6)||(b==5);
}
```

The **switchpressed** function waits for the switch to be pressed, and **switchreleased** function waits for the already-pressed switch to be released. Both functions maintain an internal counter from 1 to 6. As the switch is pressed or released, the value of the counter is picked up. Based on these counter values, LEDs are lit up, depending on number and color.

Project 13
RGB Tic-Tac-Toe

This is a modern version of the classic tic-tac-toe game, also known in some parts of the world as the naughts and crosses game, except this is not played on paper. The naughts and crosses are represented by user-selectable colors. The system offers several colors, and a user can select a color to represent the naught (or cross). Similarly, the other user can select any other available color. Once the colors are chosen, the switches on the board allow the user to position his/her naught (or cross) on any unoccupied position on the 3 × 3 board. The system uses nine integrated RGB LEDs in a 3 × 3 configuration. In this project, instead of Charlieplexing, traditional multiplexing is used to control the RGB LEDs. The bottom illustration shows the block diagram of the system. The system also has a buzzer to indicate when a player wins the game. The microcontroller recognizes when three LEDs in a row, column, or diagonal have the same color and thus terminates the game. A new game can be played after a previous game is concluded (won or lost) or if it ends in a tie.

Design Specifications

The aim of this project is to provide the classic tic-tac-toe game in a modern electronic version using RGB LEDs. A total of ten RGB LEDs are used in the system. One of the RGB LED is used to indicate the turn of the naught or the cross to play. The rest of the nine LEDs are arranged in a 3 × 3 physical configuration for the game. The system has several switches to navigate the naughts and crosses, and to position them at the selected spot.

It also has a buzzer to indicate the winner. The system has an on-board linear regulator and requires external batteries for operation.

Design Description

Until now, all the projects in this chapter used Charlieplexing to control a large number of LEDs with fewer I/O pins. This project demonstrates the concept of multiplexing and implements a traditional two-player game of tic-tac-toe. Figure 3-20 shows the schematic diagram of the project. The voltage regulator used is again LM2940 with 5V output, so input voltage can vary from around 6V to 20V. Diode D1 is a Schottky diode (IN5819) used as a protection diode, as explained earlier. Capacitor C3 is used to filter the spikes and unwanted noise in the power supply. C2 is used to stabilize the output of LM2940. C1 and C4 are soldered near the supply pins of the microcontroller to further decouple the noise arising in the circuit. The microcontroller used is

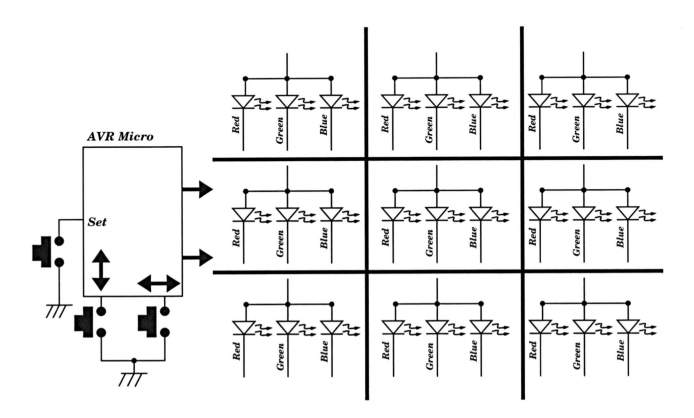

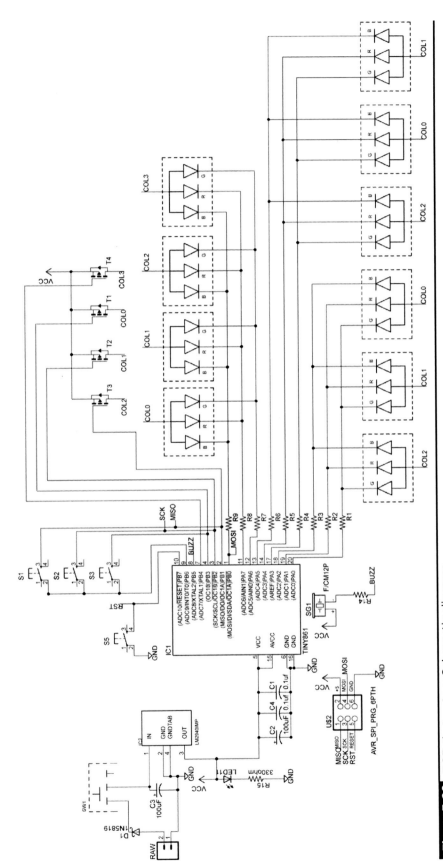

Figure 3-20 RGB tic-tac-toe: Schematic diagram

the ATtiny861 microcontroller. LEDs have been multiplexed and divided into four columns. COL0 is the first column of the 3 × 3 matrix, COL1 is the second, and COL3 is the third. Each of the three columns has three RGB LEDs. COL4 only has one RGB LED, and it is used to denote the turn during the game. T1, T2, T3, and T4 are N-MOSFETs (NDS355) and act as current-sourcing drivers for the four columns. Three switches have again been connected in a similar way as in the geek clock. They have been multiplexed with the column lines. SG1 is a buzzer used for denoting some status of the game.

The source code for this project is the most advanced of the ones used thus far. It uses software PWM on all the LEDs. Nine levels of intensity are available on each color component, so 9 × 9 × 9 colors can be generated. Out of these, 16 contrasting ones are stored in the program. As the circuit is switched on, two players select the colors they want to play with. After that, both players play the game placing their respective color dots on their position of choice through switches. The program checks for all the win conditions after every placement by either of the two players. If a player wins, the three LEDs that have formed the winning pattern start blinking and a buzzer is played for half a second. After the user presses the restart switch, the game continues with the same selection of colors but with the second player getting the first move this time, and so on.

Fabrication

The board layout in EAGLE, along with the schematic, can be downloaded from www.avrgenius.com/tinyavr1.

The board is routed in the solder layer with a few jumpers in the component layer. The component and solder sides of the soldered board are shown in Figure 3-21 and Figure 3-22, respectively.

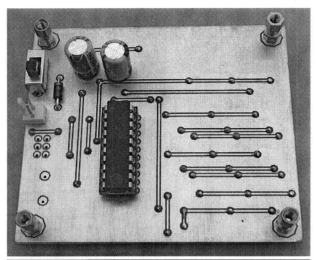

Figure 3-21 Tic-tac-toe: Component layout

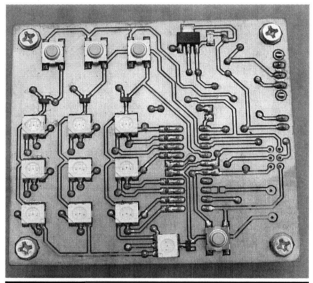

Figure 3-22 Tic-tac-toe: Solder side

Design Code

The compiled source code, along with the MAKEFILE, can be downloaded from www.avrgenius.com/tinyavr1.

The code runs at a clock frequency of 8 MHz. The controller is programmed using STK500 in ISP programming mode. The important sections of the code are shown next.

```
while(1)
{
 playerturn(b1,r1,g1,1);
 checkwin();
 playerturn(b2,r2,g2,2);
 checkwin();
}
```

This is the main infinite loop of the program. The first three arguments of the **playerturn** function denote the intensity levels of the blue, red, and green components, respectively. The fourth argument denotes the player number. The **playerturn** function is responsible for getting a player's color fixed on the LED of his choice but not on the ones that have previously been occupied by the same or another player. **Checkwin** checks all the winning conditions for both the players by using the magic square concept. A magic square is a square where the sum of each row, column, and diagonal is the same. In the source code, a particular number is assigned to each position of the LED, which is the same as what would appear on the magic square at that particular position. As the players fix their colors on the LEDs, the numbers corresponding to those positions get filled in their respective buffers. If any player has occupied the LEDs, the sum of any three of which is equal to the corresponding sum associated magic square, it means that player has won. Prior to this loop, the **start** function allows both players to pick a color of their choice for playing.

Conclusion

In this chapter, we have learned about newer types of LED configurations and how to control them using multiplexing and Charlieplexing techniques. We have also discussed in detail the intensity control of LEDs using hardware PWM features of the tinyAVR microcontrollers. Another important set of projects involved using the eight-pin microcontrollers to control a two-and-a-half-digit seven-segment display using the Charlieplexed techniques. This configuration allowed us to design several projects, which can be further modified for newer applications. In the next chapter, we take a detour to another fascinating and commonly used output device: graphics liquid crystal display (GLCD). Several projects will be illustrated using GLCD.

CHAPTER 4

Graphics LCD Projects

IN THE LAST CHAPTER, we looked at ways of interfacing LED displays in various applications. In this chapter, we look at an alternative, but equally (if not more) popular, display technology, the liquid crystal display (LCD). There are several differences between LED- and LCD-based displays. The first difference is that, unlike LEDs, which are available as a single element that can be used for displaying binary information, a single LCD element is not available. More importantly, an LED generates light whereas LCD manipulates ambient light. Due to this, an LED display is suitable for use in low light conditions (such as at night), whereas an LCD is perfect for use during the day where there is ample ambient light available. However, the designer of a gadget must make a choice between an LED and an LCD. If an LCD display is to be used and it is likely to be used in low ambient lighting conditions, then some sort of backlight must be provided. Interestingly, the backlight is sourced by LEDs. Shown in the photo are a few popular LCDs.

LCDs are of a few types: reflective, transmissive, and transreflective. In the next section we look at the working of these devices and subsequently use them in several projects.

Principle of Operation

As mentioned, there are three types of LCDs: reflective, transmissive, and transreflective. The

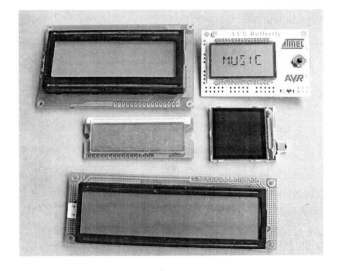

various layers of a reflective LCD are shown in Figure 4-1. At the bottom it contains a mirror that reflects the incoming light from the top. It has polarizing film layers (B and F). Between these films is a layer of liquid crystal (D) sandwiched between two layers of electrodes (C and E). In the unpowered state, the liquid crystal allows light coming from the top to pass through, bounce back from the mirror, and go back through the top. However, on application of an electric signal between the electrodes, the liquid crystals, as shown in Figure 4-2, are so oriented that they block the passage of light, and to an observer, a black rectangle marked on the upper electrode is visible. A reflective LCD would need ambient light to display anything. If display readability at all times is required, then some sort of backlight is required. White LEDs are quite popular for

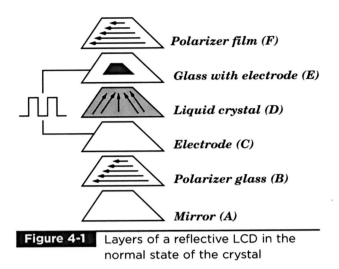

Figure 4-1   Layers of a reflective LCD in the normal state of the crystal

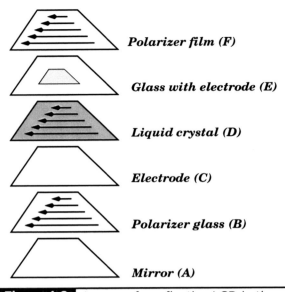

Figure 4-2   Layers of a reflective LCD in the polarized state of the crystal

providing backlight for reflective LCDs. These are put above and/or beside the LCD to provide uniform illumination.

In the transmissive LCD, the bottom polarizer is transparent and does not have any mirror. They necessarily require backlight for operation. Transmissive LCDs work very well under low ambient lighting conditions with their backlight continuously on. The transreflective LCD uses a partially reflective rear polarizer. It has a source of backlight, too, and can be turned on when the ambient light is insufficient.

The next illustration shows the block diagram of a typical LCD controller. Apart from the matrix of liquid crystal elements, the controller has memory to store the information that needs to be displayed on the LCD. The number of memory bits is equal to the number of liquid crystal elements. These bits may be arranged in bytes for ease of access. Also, to polarize the liquid crystal, the controller needs high voltage bias in the range of 5V to 10V. Usually, the bias voltage (V(LCD)) generator is built into the controller. To communicate with the external world, the controller would have some sort of communication bus. It could be a byte-wide parallel bus or a more frugal SPI or I2C (inter-IC communication) interface. An external microcontroller would communicate with the LCD

controller through the communication bus to send display data, etc. The LCD controller would receive the data, store it in its internal memory, and use it to display the information appropriately.

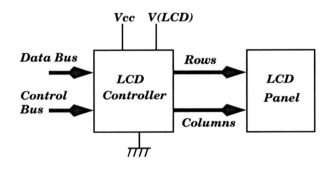

The Nokia 3310 is a popular mobile phone, and due to its popularity, many of its components are available for replacement and repair. The graphics LCD on the Nokia 3310 phone is easily available—probably not the original displays, but compatible and inexpensive "Made in China" displays are easily available. In the next section, we describe this immensely popular display that can be easily integrated in small embedded projects.

Nokia 3310 GLCD

The Nokia 3310 LCD is a small graphical LCD (GLCD), suitable for various projects related to embedded systems. It is widely used because it can be interfaced easily with most of the microcontroller families and is easily available. The display is 38 × 35 mm, with an active display surface of 30 × 22 mm and an 84 × 48 pixel resolution. The display comes with an LCD controller/driver, the PCD8544, designed to drive a graphic display of 48 rows and 84 columns. The display is easy to interface using standard SPI communication. Some of its other important features include the following:

- Only one external component is needed, a capacitor of 1–10μF value between VOUT and GND

- Logic supply voltage range: 2.7 to 3.3V

- Low power consumption, suitable for battery-operated systems

- Temperature range: –25 to +70°C

The datasheet of the LCD can be downloaded from www.avrgenius.com/tinyavr1.

Interfacing the Nokia 3310

The Nokia 3310 works with the SPI interface, which exists in many tinyAVR microcontrollers but is absent in many other Tiny controllers. Also, the SPI is often busy interacting with other devices. But this does not mean that this LCD cannot be used with these devices. We can interface this LCD by implementing the software SPI interface, also called "bit banging," that can be used on any I/O pin of the microcontroller. For this purpose, we require a minimum of four I/O pins of the controller. The following illustration shows the pin-out details of the LCD display. A detailed description of this pin-out is discussed next.

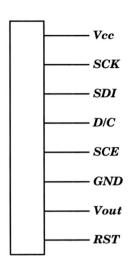

The LCD pin-out consists of eight pins as follows:

- **VCC**   Input voltage connected to regulated voltage supply (2.7 to 3.3V).

- **SCK**   Input for serial clock signal (0.0 to 4.0 MB/sec). Connected to I/O pin.

- **SDI**   Serial data input. Connected to I/O pin.

- **D/C**   Data/Command mode select input. Connected to I/O pin.

- **SCE**   Chip select. This pin can be connected to an I/O pin of a microcontroller or can be grounded (to always select LCD). This depends upon the requirement of the project.

- **GND**   Ground.

- **VOUT**   VLCD. This pin is connected to GND via a 10μF capacitor.

- **RST**   Reset pin of the PCD8455 controller. Connected to I/O pin.

Functional Description of the PCD8455

The PCD8455 is a low-power CMOS LCD controller/driver, designed to drive a graphic display of 48 rows and 84 columns. All necessary functions for the display are provided in a single chip, including on-chip generation of LCD supply and bias voltages, resulting in a minimum number of external components and low power consumption. Figure 4-3 shows the block diagram of the PCD8455 controller.

Some important features of this controller are:

- **Address counter**   The address counter contains the addresses of the display data RAM for addressing a single column of 8 pixels. The X-addresses, 0 to 83, and the Y-addresses, 0 to 5, are set separately.

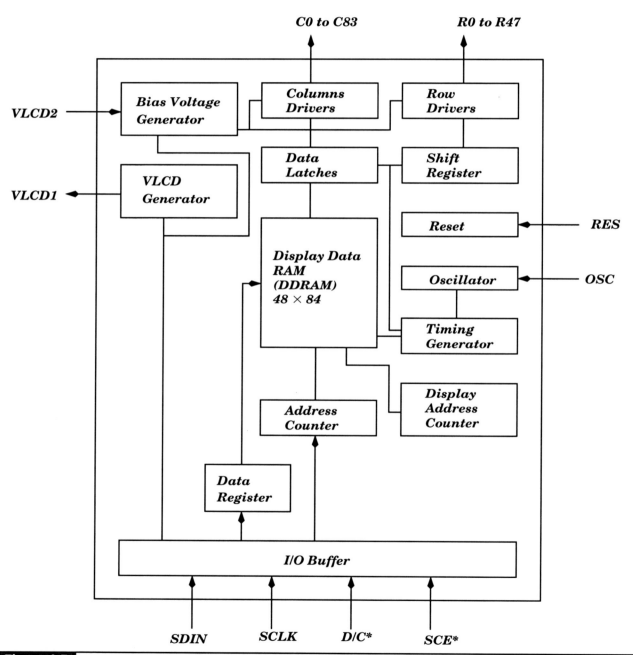

Figure 4-3   Block diagram of the PCD8455

- **Display data RAM** The DDRAM is a 48 × 84–bit static RAM that stores the display data. The RAM is divided into six pages of 84 bytes (6 × 8 × 84 bits). Each page is addressed by a single Y address, and the individual columns of each page are addressed by a single X address.

- **Instructions** The instruction format is divided into two modes. If D/C (mode select input) is set LOW, the current byte is interpreted as a command byte; otherwise, if the pin is HIGH, the current byte is stored as a data byte in the display data RAM. After every data byte, the address counter is incremented automatically. D/C signal is read after the last bit of the byte has been transmitted, that is, at the eight SCLK pulse. When SCE is kept high, the SLCK signal is ignored and the serial interface is initialized. Data (SDIN) is sampled at the positive edge of the clock pulse. If the SCE signal stays low even after transmission of the last bit of the command/data byte, the controller is set to accept the next byte, assuming the incoming bit to be the seventh bit of the next byte. An active low RESET pulse resets the controller by interrupting the transmission and clearing all the registers. If SCE is LOW after the positive edge of the RESET pulse, the controller is ready to accept the next byte.

Design Code

The required C library to interface the LCD can be downloaded from www.avrgenius.com/tinyavr1.

These functions are integrated with the source codes of all the projects that have been used in this chapter.

```
void clockdata(char bits_in)
{
    int bitcnt;
    for (bitcnt=8; bitcnt>0; bitcnt--)
    {
        LCD_PORT = LCD_PORT&(~(1<<SCK));
        // Set Clock Idle level LOW.
```

```
        if ((bits_in&0x80)==0x80)
    {LCD_PORT |=1<<SDIN;}
        else {LCD_PORT &= ~(1<<SDIN);}
        LCD_PORT |=1<<SCK;
    // Data is clocked on the rising
    // edge of SCK.
        bits_in=bits_in<<1;
    // Logical shift data by 1 bit
    // left.
    }
}
```

This is the routine used to transmit a byte of data from the microcontroller to the LCD driver. A loop is used to clock data, bit by bit, to the LCD controller. To transmit a bit, first **SCK** is held **LOW** and then the bit to be transmitted is brought on the **SDIN** pin connected to the LCD. After setting the bit, **SCK** is held **HIGH** as data is transmitted on the rising edge of the clock pulse. This process is repeated eight times by the loop to transmit the required byte. Note that we have grounded the **SCE** pin of the LCD.

```
void writecom(char command_in)
{
LCD_PORT = LCD_PORT&(~(1<<D_C));
    // Select Command register.
clockdata(command_in);
    // Clock in command bits.
}
void writedata(char data_in)
{
LCD_PORT = LCD_PORT|(1<<D_C);
    //Select data register.
clockdata(data_in);
    // Clock in data bits.
}
```

These routines use the **clockdata** function to transmit either a command byte (by holding D/C mode to select pin low), as in **writecom** function, or a data byte, as shown in **writedata** function. Now we know how to transfer a data byte to an LCD, but in order to use an LCD, we need to initialize it according to the described procedure in its datasheet. This is shown here:

```
void initlcd(void)
{
        LCD_DDR |=
    1<<SCK|1<<SDIN|1<<D_C|1<<RESET;
        LCD_PORT = LCD_PORT|1<<RESET;
        LCD_PORT =
    LCD_PORT&(~(1<<RESET));
        _delay_ms(200);
        LCD_PORT = LCD_PORT|1<<RESET;
        writecom(0x21);
            // Activate Chip and H=1.
        writecom(0xD3);
            // Set LCD Voltage to about
            // 9V.
        writecom(0x13);
            // Adjust voltage bias.
        writecom(0x20);
            // Horizontal addressing
            // and H=0.
        writecom(0x09);
            // Activate all segments.
        clearram();
            // Erase all pixel on the
                DDRAM.
        writecom(0x08);
            // Blank the Display.
        writecom(0x0C);
            // Display Normal.

}
```

In this routine, first we declare all four I/O pins as outputs using the corresponding DDR register. Then we initialize the LCD by sending a RESET pulse of 200 ms. This clears any previous setting in the LCD. Then the LCD is initialized according to our requirements by sending a series of command words. First, the LCD is activated with an extended instruction set (H = 1). The next command byte is used to set the operating LCD voltage by software. Using this feature, one can set the contrast of the LCD in use. This is done by calculating the eight-bit value to be transmitted using the relation described in the datasheet of the Nokia 3310 LCD. Then the voltage bias is adjusted

by selecting a multiplexing rate of 1:48. The next command byte is used to select horizontal addressing and to make H = 0 so as to use commands like "set X address" and "set Y address." To activate the display, first a command byte is sent to activate all segments in the display, the next byte is used to blank the display, and the last byte is sent to use the display in normal mode. This routine should be called whenever one wants to use the display in normal mode.

Glitches Observed in Certain Displays

In certain displays, sometimes we have to confront a problem of a broken first page in the 48 × 84–pixel screen. A usual Nokia 3310 contains six (zero to five) pages, but in some cases, the topmost page with Yaddress of 0 is broken. Only five pixels of each column of this page are visible. On the lower side, we get a new page with three pixels in each column. This new page is addressed by Yaddress equal to 6. If we want to do away with this arrangement, we can change the initializing source code of the LCD. In the **initlcd** function, writing the command **writecom(0x45)** after the bias voltage adjustment would cause the pages to readjust and shift vertically upwards by five pixels. This completely hides the broken top page and makes the lower page completely visible. In this case, Yaddresses vary from 1 to 6.

The following illustrations show the Nokia display and keypad, the mounting PCB, and the final form of the display soldered onto a custom PCB.

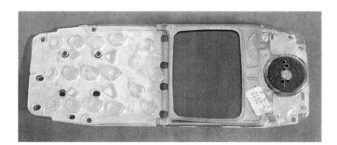

Project 14
Temperature Plotter

This project uses a temperature sensor and displays the ambient temperature in degrees Celsius and Fahrenheit on the display. It also displays the minimum and maximum temperatures recorded by

the sensor. The illustration below shows the block diagram of the project. The Nokia 3310 display is used to display the readings. A switch on the circuit board is used to toggle between two screens. One screen shows the numerical value of the temperature alternately in degrees Celsius and

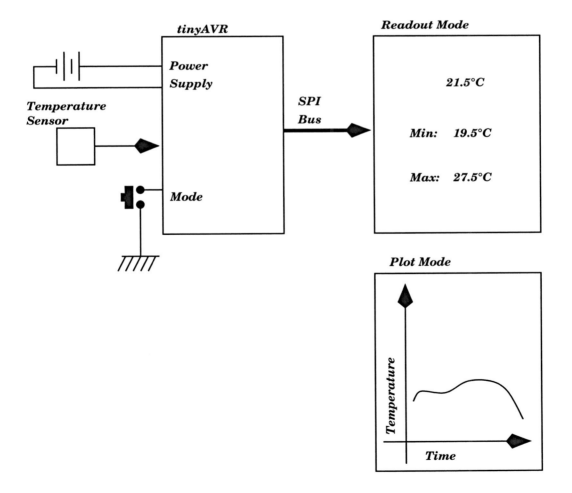

Fahrenheit. On the other screen, the system plots the variation of the temperature as a function of time. The system is battery powered.

Design Specifications

The objective of the project is to design a temperature display system to show the ambient temperature value in degrees Celsius and Fahrenheit, together with minimum and maximum values. The system should also be capable of recording temperature variation as a function of time. The system should be battery driven so that it is portable and can be installed anywhere.

Design Description

Figure 4-4 shows the schematic diagram of the project. Since the project uses the Nokia display, it requires a supply voltage between 2.7V and 3.3V. The power supply is based on the TPS61070 step-up DC-DC converter to provide 3.3V and, therefore, can be powered with a single 1.5V battery. The battery is connected to the SL3 connector. Since there is no polarity protection, utmost care should be taken to ensure that the battery is connected properly. The Nokia display is connected using the SPI bus through SL1.

The most important component of the system is the temperature sensor. There are several options for this component: a thermistor, a thermocouple, or a bandgap semiconductor sensor. A semiconductor sensor is the easiest to use.

There are also various semiconductor temperature sensors. Some provide direct analog voltage output proportional to the temperature, while others provide a digital value in degrees Celsius or Fahrenheit directly. We used the DS1820 one-wire temperature sensor, which converts the temperature to a nine-bit digital number representing the temperature in degrees Celsius or Fahrenheit. The temperature reading has a resolution of 0.5 degrees Celsius or 0.9 degrees

Fahrenheit in the range of –55°C to +125°C or –67°F to +257°F. The sensor takes 200 ms for conversion. For more details, please refer to the DS1820 datasheet on our website at www.avrgenius.com/tinyavr1.

The converted data can be read out of a single-wire interface. In the schematic diagram, SL2 represents the DS1820 sensor. The circuit also has four switches—S1 through S4—but for this project, only one switch is used. The rest of the switches are used in other projects in this chapter. The circuit uses the Tiny44 microcontroller in 14-pin SMD version with 4KB of program memory. Upon power or reset, the microcontroller initializes the display and queries the DS1820 sensor and displays the temperature in Celsius and Fahrenheit on the display. It also maintains the observed minimum and maximum temperature values. The user can press the switch at any time, and the system will switch to a different display screen and mode where the time versus temperature readings are plotted. The system takes temperature readings continuously but only stores one temperature reading every ten minutes and plots the graph on the display. The system can store a maximum of 40 readings and, hence, can show the variation in temperature for the last 400 minutes. The buffer for storing the reading is continuously shifted to accommodate new values, thereby flushing out the earlier ones.

Fabrication

The board layout in EAGLE, along with the schematic, can be downloaded from www.avrgenius.com/tinyavr1.

The board is routed in the component (top) layer, with few jumpers in the solder (bottom) layer. The component side and solder side of the board are shown in Figures 4-5 and 4-6, respectively. Soldering the TPS61070 IC is critical and should be done carefully. Start by soldering the regulator and other associated components. The

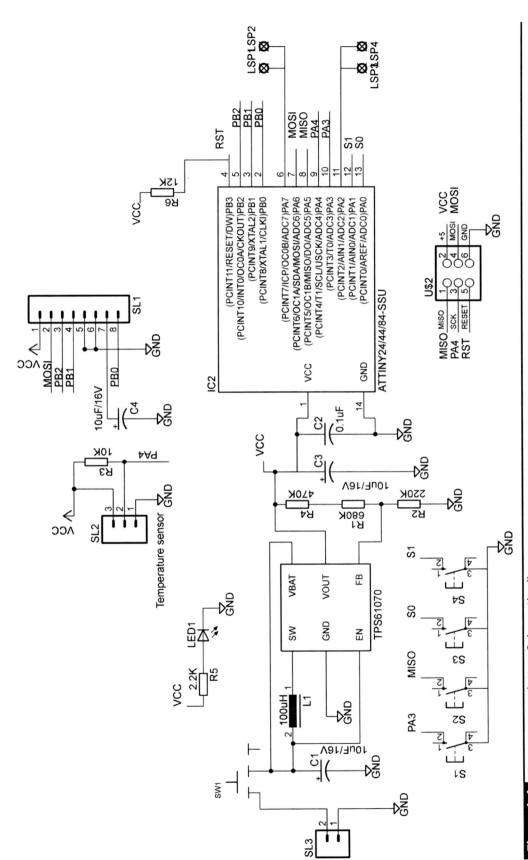

Figure 4-4 Temperature plotter: Schematic diagram

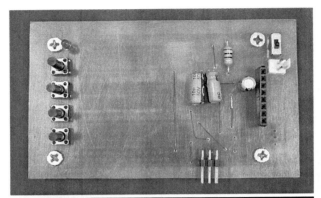

Figure 4-5 Temperature plotter: Component side

Figure 4-6 Temperature plotter: Solder side

Figure 4-7 Temperature plotter display in readout mode

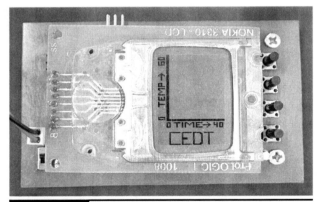

Figure 4-8 Temperature plotter display in graph mode

output of the TPS61070 should be tested before soldering the rest of the components. The plotter displays in different modes are shown in Figures 4-7 and 4-8.

Design Code

The compiled source code, along with the MAKEFILE, can be downloaded from www.avrgenius.com/tinyavr1.

The code runs at a clock frequency of 1 MHz. The controller is programmed using STK500 in ISP programming mode. The temperature sensor DS18S20 performs read and write functions through a Dallas one-wire interface. This interface has been implemented in the software. You can refer to the datasheet of the temperature sensor to

get the gist of the various commands. The important sections of the code are explained here:

```
int ds1820_read(void)
{
  char busy=0;
  unsigned char temp1,temp2;
  int result;
  onewire_reset();
  onewire_write(0xCC);//Skip Rom Command
```

```
onewire_write(0x44);
    //Temperature Conversion command
while (busy == 0)
busy = onewire_read();
onewire_reset();
onewire_write(0xCC);//Skip Rom command
onewire_write(0xBE);
    //Read ScratchPad Command
temp1 = onewire_read();
temp2 = onewire_read();
onewire_reset();
result = temp1*5;
    // 0.5 deg C resolution
//result is ten times actual
//temperature
return result;
}
```

The **ds1820_read** function reads the DS1820 and returns a value that is ten times the actual temperature in degrees Celsius, after performing the necessary scaling. The main infinite loop of the program operates in two modes. In the first mode, it displays the present temperature, along with the maximum and minimum values, in both degrees Celsius and Fahrenheit. In the other mode, it displays the variation of temperature in the form of a graph. The graph is drawn using the **graph1** function, which extracts the values of array **data** to plot the pixels. The function **setlcd** is used to draw the axis on the LCD screen. Switch S4 (PA1) is used to toggle the mode in this project. Shifting from plot mode to temperature mode and vice versa doesn't delete the status of the graph. Other sections of the code handle the LCD initialization and graphics.

Working

The temperature plotter is designed to operate from either one or two AA/AAA batteries. Alkaline or rechargeable batteries such as NiMH or NiCd batteries can be used. Just apply power, and the display starts showing the temperature value. Use the switch to switch between the readout and plotter modes.

Project 15
Tengu on Graphics Display

Tengu refers to certain supernatural creatures in Japanese folklore. However, Tengu is also a famous toy that interacts with the user through ambient music, sounds, or noise. This project is the audio-based toy and has nothing to do with supernatural creatures. Tengu shows a face with eyes, nose, and mouth. Depending upon the ambient noise, the Tengu eyes and facial expressions change. The illustration below shows the block diagram of the Tengu project. It uses an audio amplifier with a microphone input to sense the ambient sounds. It uses the Nokia graphics LCD to display the face. The tinyAVR

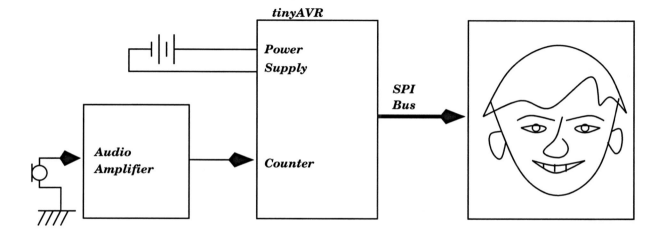

microcontroller samples the sound input and changes the facial expressions appropriately. The circuit is battery powered and can be carried around.

Design Specifications

The objective of the project was to design a battery-powered Tengu clone that would respond

to the ambient sounds, music, and noise by changing its facial expressions.

Design Description

Figure 4-9 shows the circuit schematic of Tengu. The circuit is operated with a 9V battery (even four 1.5V alkaline batteries would work fine). The DC input voltage is connected to the circuit through the SL2 connector. The series diode D2

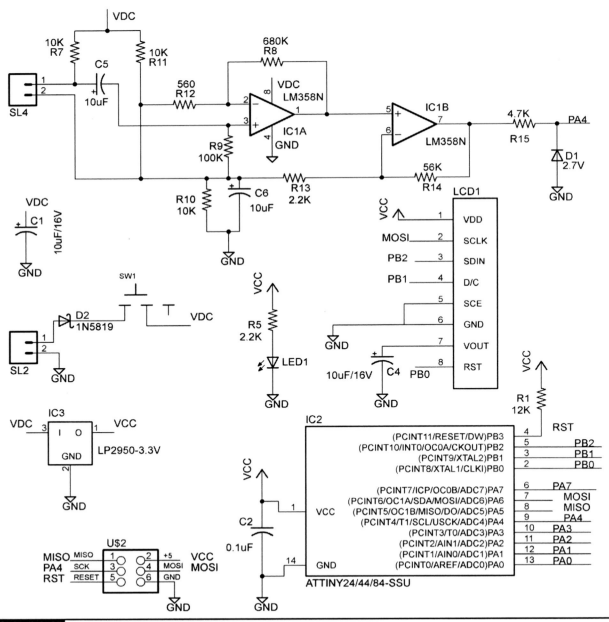

Figure 4-9   Tengu: Schematic diagram

provides reverse polarity protection in case the battery is connected incorrectly. The DC voltage provides the supply voltage to the dual op-amp LM358. It is also connected to the input of an LDO LP2950-3.3V that provides a supply voltage Vcc of 3.3V for the microcontroller and the Nokia display. The display is connected to the circuit through the LCD1 connector. The microphone is connected to the circuit through the SL4 connector. A condenser microphone is connected to the SL4 connector (polarity of the microphone is important when connecting the mic to the connector). Resistor R7 biases the condenser mic, and the analog signal produced by the mic (in response to the audio sound) is capacitively coupled to the op-amp circuit through a 10uF capacitor (C5). The amplifier is set up as a noninverting, high-gain AC amplifier. The junction of resistances R10 and R11 provides a DC bias equal to half the DC input voltage. Since the resistor R10 is shunted with a capacitor C6, the junction of R10 and R11 acts like AC ground. The AC signal rides the DC bias and is connected to the amplifier input on pin 3.

The output of the first stage of the amplifier is further enhanced using the second stage which uses the other half of the dual op-amp, which is configured as a noninverting amplifier. The output at pin 7 is passed through a resistor R15 and a Zener diode D1 (2.7V) so that the output at the Zener diode cathode is clipped to 2.7V. This output is connected to the input pin PA4 of the Tiny44 microcontroller. The gain of the amplifier is so large (25,000) that the microphone output is actually converted into a square wave. The frequency of the square wave is the same as that of the audio signal. Thus, the task of the microcontroller is to measure the frequency of this signal and to then change the facial expressions on the graphics display appropriately. The microcontroller continuously measures the frequency of the audio signal every second.

The op-amp is directly powered by the battery voltage. LM358 requires a supply voltage of at least 5V; hence, the circuit must be powered with either four 1.5V batteries or a 9V battery. The microcontroller and the display are powered by the output of the LP2950-3.3 regulator. This regulator provides 3.3V output voltage and is necessitated by the requirements of the Nokia 3310 display.

Fabrication

The board layout in EAGLE, along with the schematic, can be downloaded from www.avrgenius.com/tinyavr1.

The board is routed in the component (top) layer with a few jumpers in the solder (bottom) layer. The component side and solder side of the board are shown in Figures 4-10 and 4-11, respectively. A working demonstration is shown in Figure 4-12.

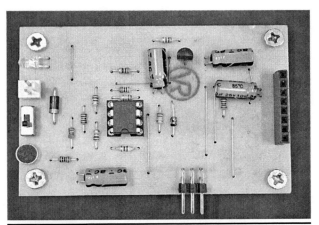

Figure 4-10 Tengu: Component side

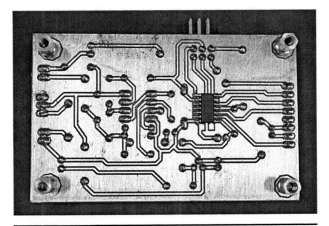

Figure 4-11 Tengu: Solder side

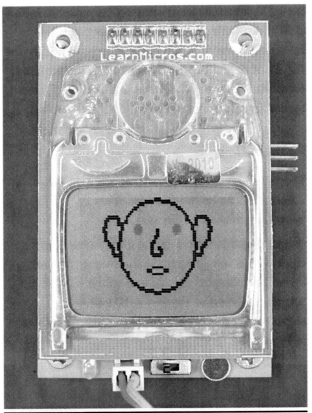

Figure 4-12 Tengu: Working demonstration

Design Code

The compiled source code, along with the MAKEFILE, can be downloaded from www.avrgenius.com/tinyavr1.

The code runs at a clock frequency of 8 MHz. The controller is programmed using STK500 in ISP programming mode. The important sections of the code are explained here:

```
void check(void)
{
  TIFR0 = 1<<TOV0;
  TCNT0 = 255-125;
  TCCR0B = 1<<CS00|1<<CS01;
  count = 0;
  freq = 0;
  stop_check=0;
  while(count<200)
  {

    while((PINA&(1<<4))&&(stop_check==0)
```

```
    );
    while((!(PINA&(1<<4)))&&(stop_check
      ==0));
    if(PINA&(1<<4))
    {
      freq++;
    }
  }
  freq = freq*5;
  if(freq>=1900)
  face = 9;
  else if(freq>=1700)
  face = 7;
  else if(freq>=1500)
  face = 6;
  else if(freq>=1200)
  face = 1;
  else if(freq>=900)
  face = 0;
  else if(freq>=650)
  face = 5;
  else if(freq>=350)
  face = 2;
  else if(freq>=200)
  face = 4;
  else if(freq<200)
  face = 3;
  TCCR0B = 0x00; //stop timer
}
```

This is the function that measures the frequency of the input signal on the PA4 pin of the microcontroller. It initializes Timer0 in such a way that its overflow interrupt occurs every 1 ms. ISR for Timer0 increments a variable **count** till 200, that is, interval of 200 ms. During this interval, each logic change on the PA4 pin is recorded in the variable freq. This is then multiplied by 5 to get the number of logical changes in 1 second and, hence, gives the frequency of the input signal. Using this frequency, a particular face number is assigned to the **face** variable. This assignment has been done using a trial-and-error process. The function **putface** uses this variable to display the desired face on the LCD by extracting the display bytes from the program memory of the microcontroller. It also displays blinking eyes by calling the **puteye** function.

The main function of the code first initializes the LCD and then calls **check** and **putface** functions alternately in an infinite loop.

Working

To use Tengu, you need to power the circuit board and then place it so that the microphone is not hindered in any way. You would see the face on the display and when there is some sound, the eyes would blink and the mouth patterns would change. If you play loud music, the mouth patterns would change as if Tengu is singing.

Project 16
Game of Life

Game of Life is a mathematical simulation devised by John Conway. It is an attempt to model real-life processes using simple rules. It's a zero player game, that is, once the initial state is set, it does not require any intervention to proceed. Details of the game are available on the Internet, and you are strongly advised to read them. This project allows a user to simulate the Game of Life on a Nokia display using a tinyAVR microcontroller. The original game developed by John Conway uses a

two-dimensional grid of infinite dimensions, but this project limits the size to 16 × 16 elements. The user can set up the initial state with the help of switches, as seen in the illustration below. Once the initial state is fixed, the game can run to show the evolution.

Design Specifications

The objective of the project is to create the Game of Life simulation on a small microcontroller with graphics display and allow the user to create any initial state to run the simulation. The system has four switches to "turn on" any required element in the 16 × 16 grid and then to run the simulation. The system is battery powered for portability.

Design Description

The circuit for this project is the same as for the temperature plotter project. In the temperature plotter project, only one of four available switches was used, but in this project, all the four switches are used. The purpose of two of the four switches is for left/right and up/down movement of the cursor; these are called the arrow keys. The third switch is for toggling the state of the element, and the fourth switch is for running the simulation. A demonstration of the output display on the LCD is shown in Figure 4-13.

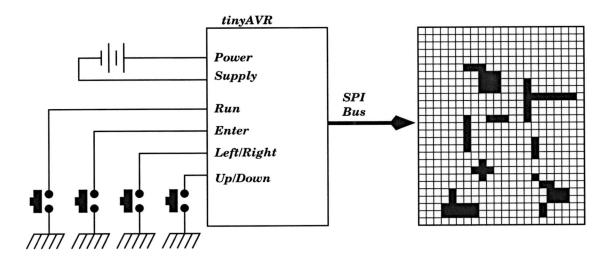

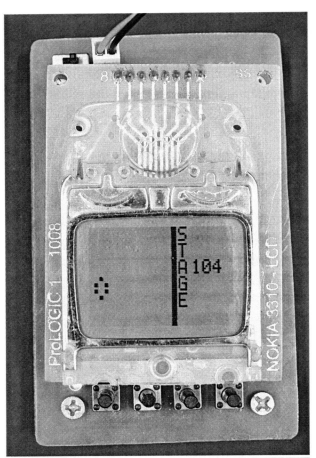

Figure 4-13 Output display for the Game of Life

Design Code

The compiled source code, along with the MAKEFILE, can be downloaded from www.avrgenius.com/tinyavr1.

The code runs at a clock frequency of 8 MHz. The controller is programmed using STK500 in ISP programming mode. The important sections of the code are explained here:

```
void place(void)
{
 row=0;
 column=0;
 while(1)
 {
  if(!(PINA&0b00000001))
  {
   _delay_ms(30);
   while(!(PINA&0b00000001));
```

```
_delay_ms(30)
TIMSK1 = 0x00;
if(led[0][row]&(1<<column))
{
 pix_light(row,column,1);
}
else if(!(led[0][row]&(1<<column)))
{
 pix_light(row,column,0);
}
 row++;
 if(row==(ROWMAX+1))
 row = 0;
 TIMSK1 = 0x01;
}
if(!(PINA&0b00000010))
{
 _delay_ms(30);
 while(!(PINA&0b00000010));
 _delay_ms(30);
 TIMSK1 = 0x00;
 if(led[0][row]&(1<<column))
 {
  pix_light(row,column,1);
 }
 else if(!(led[0][row]&(1<<column)))
 {
  pix_light(row,column,0);
 }
 column++;
 if(column==(COLMAX+1))
 column = 0;
 TIMSK1 = 0x01;
}
if(!(PINA&0b00100000))
{
 _delay_ms(50);
 while(!(PINA&0b00100000));
 _delay_ms(50);
 TIMSK1 = 0x00;
 if(led[0][row]&(1<<column))
 led[0][row]&=~(1<<column);
 else led[0][row]|=1<<column;
 if(led[0][row]&(1<<column))
 {
  pix_light(row,column,1);
 }
 else if(!(led[0][row]&(1<<column)))
 {
  pix_light(row,column,0);
```

```
      }
      row++;
      if(row==(ROWMAX+1))
      row = 0;
      TIMSK1 = 0x01;
    }
    if(!(PINA&0b00001000))
    {
      _delay_ms(30);
      while(!(PINA&0b00001000));
      _delay_ms(30);
      TIMSK1 = 0x00;
      if(led[0][row]&(1<<column)
      {
        pix_light(row,column,1);
      }
      else if(!(led[0][row]&(1<<column)))
      {
        pix_light(row,column,0);
      }
      TIMSK1 = 0x01;
      break;
    }
  }
}
```

This function is used to place the initial population of alive cells. Each cell on the LCD is composed of nine pixels. The top left 2 × 2 pixels light up for active cells, whereas the remaining five pixels are always off. This prevents the merging of adjacent pixels. The total grid is a 16 × 16 matrix of such cells. To optimize the RAM usage, we have used different int arrays of length 16 to denote the entire grid. Each variable of the array denotes one row, whereas its 16 bits denote 16 columns. The function uses four switches to set the initial population. Two switches are used to move the cursor row-wise or column-wise, the third is used to toggle the state of the cell at the present cursor location, and the last is used to start the game.

```
while(1)
{
  clearram();
  stage=0;
```

```
for(unsigned char j=0;j<=5;j++)
{
  cursorxy(48,j);
  writedata(0xFF);
  writedata(0xFF);
  writedata(0xFF);
}
TIMSK1 = 0x01;
  //Overflow Interrupt Vector
place();
TIMSK1 = 0x00;
  //Overflow interrupt Disabled
cursorxy(52,0);
putcharacter('S');
cursorxy(52,1);
putcharacter('T');
cursorxy(52,2);
putcharacter('A');
cursorxy(52,3);
putcharacter('G');
cursorxy(52,4);
putcharacter('E');
generation=0;
generation1=1;
while(1)
{
  for(int row=0;row<=ROWMAX;row++)
  {
    for(int col=0;col<=COLMAX;col++)
    {
      if(led[generation][row]&(1<<col))
      led[generation1][row] |=1<<col;
      else led[generation1][row]
        &=~(1<<col);
      if(led[generation][row]&(1<<col))
        //on
      {
        check_neighbors(generation,row,
          col);
        if(alive>3||alive<2)
        {
          led[generation1][row] &= ~(1<<col);
          pix_light(row,col,0);
        }
      }
      else
      if(!(led[generation][row]&(1<<col)))
      {
        check_neighbors(generation,row,
```

(continued on next page)

```
        col);
      if(alive == 3)
      {
        led[generation1][row]  |= 1<<col;
        pix_light(row,col,1);
      }
     }
    }
   }
generation = (generation==0)?1:0;
generation1 = (generation==0)?1:0;
//rules of game of life
stage++;
stage1=stage;
cursorxy(60,2);putcharacter (stage1/
  100+48);
stage1 = stage1%100;
putcharacter(stage1/10+48);
stage1 = stage1%10;
putcharacter(stage1+48);
_delay_ms(500);
flag=0;
for(unsigned char i=0;i<=15;i++)
{
 if(!(actual[i]==0))
  {
  flag=1;
  break;
  }
 }
if((flag==0)||(stage==255))
{
  clearram();
  for(unsigned char i=0;i<=15;i++)
  {
   led[0][i]=0;
   led[1][i]=0;
   actual[i]=0;
   flag=0;
  }
  cursorxy(12,2);
  putstr("GAME OVER");
  while(PINA&0x01);
  _delay_ms(30);
  while(!(PINA&0x01));
  _delay_ms(30);
  break;
 }
}
}
```

This is the main infinite loop of the program. It works on two alternating generations of populations. While checking a particular generation, it checks the next state of each cell according to the rules of the game and updates the display. Simultaneously, it fills the next state of each cell in the appropriate location of the array **led**. It checks to see if a cell has more than three or fewer than two live neighbors (a cell in the grid can have a maximum of eight neighbors). If either of these two conditions is true, that cell is turned off (becomes dead). In case a cell has exactly three neighbors, it is kept on if already alive, or switched on if dead. After each cell has been checked, the next state becomes the present state and so on. Only two variables, **generation** and **generation1**, are required to iterate between present and next generations. The game terminates if all the cells die (are turned off) or when the value of **stage** reaches 255, whichever happens first. Variable stage represents the number of generations gone.

The other parts of the code are the initializations for various variables, the functions for interacting with the LCD, and an Interrupt Service Routine (ISR) for the overflow of TIMER1. The ISR comes into play only during the setting of the initial population to blink the cell presently pointed to by the cursor.

Working

To use the project, just apply power to the circuit board. A blinking cursor at the top left of the screen appears. Use the arrow keys to reach any required element and the toggle key to toggle the state of the element. Once the initial state of the simulation is fixed, the run switch is pressed to start the simulation. The display shows the iterations on the right side of the screen.

Project 17
Tic-Tac-Toe

As the name suggests, this project allows two people to play the popular game of tic-tac-toe (also known as naughts and crosses). The illustration below shows the block diagram of the project.

Design Specifications

The objective of the project is to provide a user interface to play the tic-tac-toe game on the Nokia graphics display. The user places the naught (or cross) using switches. The game ends either in a draw or when one of the players wins. The project keeps track of the number of games played and the individual winning score. The system is battery powered so that it is portable.

Design Description

The project uses the same circuit as in the temperature plotter project. Out of the four switches on the circuit board, this project uses three switches, labeled Up/down, Left/right, and Enter. Two output displays are shown in Figures 4-14 and 4-15.

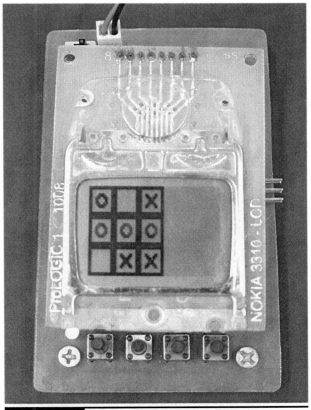

Figure 4-14 Tic-tac-toe output display

Design Code

The compiled source code, along with the MAKEFILE, can be downloaded from www.avrgenius.com/tinyavr1.

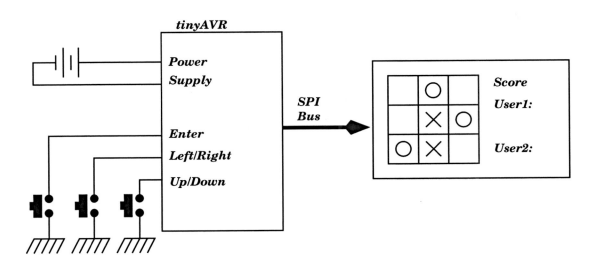

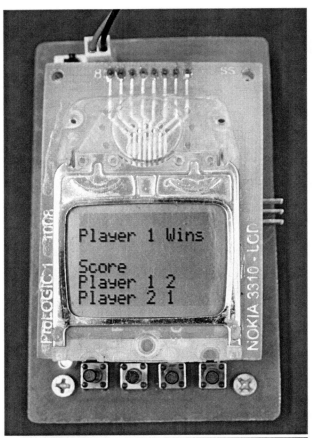

Figure 4-15 Tic-tac-toe result display

The code runs at a clock frequency of 8 MHz. The controller is programmed using STK500 in ISP programming mode. The logic behind the implementation of the game in this code is the same as that used in the RGB tic-tac-toe discussed in Project 13. The only difference is that there is no selection of color initially. One user is assigned to crosses (or X's) and the other is assigned to zeroes (O's). The bit patterns for "cross" and "zero" are stored in the program memory. Unlike the previous version, this version stores the number of wins and losses of each user until one of them wins ten times. At the end of every game, the result and total score of each user is displayed on the LCD. The important sections of the code are explained here:

```
while(1)
{
  cli();
  cursorxy(6,2);
  putstr(DISPLAY);
  while(PINA&0x01);
  _delay_ms(30);
  while(!(PINA&0x01));
  _delay_ms(30);
  sei();
  TIMSK1 = 0x01;
      //Overflow interrupt enabled
  reset();
  while(1)
  {
  playerturn(1);
  checkwin();
  if(dis1==10)
  {
    dis1=0;
    dis2=0;
    break;
  }
  playerturn(2);
  checkwin();
  if(dis2==10)
  {
    dis1=0;
    dis2=0;
    break;
  }
  }
}
```

This is the main infinite loop of the program. It puts "TIC TAC TOE" on the screen and waits for the user to press a switch on PA0 before starting the game. As the game starts, it calls **playerturn** with the player number as the argument. The **playerturn** function allows the specified player to place his or her symbol at the desired location. It then calls **checkwin** to see if the player who last placed his or her symbol won or not. Variables **dis1** and **dis2** keep track of the number of wins of each player.

Working

To use the project, apply power to the circuit. The first player gets the chance to place his or her symbol and then the other player and so on.

Project 18
Zany Clock

A microcontroller-based clock is passé. But this project is different. Instead of the usual display showing the digits on an LCD or LED display, this project shows the seconds, minutes, and hours scrolling past a marker, which is why it is titled Zany Clock! Its block diagram is shown below.

Design Specifications

The objective of this project is to design a microcontroller-based clock with an unusual display that shows the time scrolling past a marker. The project is battery operated so that it can be portable and can work even in the absence of mains power.

Design Description

Figure 4-16 shows the schematic diagram of the project. It uses a Tiny861 microcontroller and a Nokia display to show the time. The microcontroller uses an external crystal of 7.3728 MHz to generate the system clock frequency. The same system clock frequency is used to maintain and manage real time. The project is battery powered with a 9V battery (four 1.5V batteries could also be used). The battery voltage is regulated with a LP2950-3.3V regulator to power the microcontroller as well as the Nokia display. The circuit shows additional components such as an op-amp and a connector for a condenser mic, but these components are not associated with the clock project. These components are for a different project. The system has two switches, S1 and S2, which are used to set the time and to scroll the display. The clock maintains seconds, minutes, hours, and days of the week. However, only three items can be seen at any given point, due to the limited size of the display. Switch S2 is used to scroll through all these components.

Fabrication

The board layout in EAGLE, along with the schematic, can be downloaded from www.avrgenius.com/tinyavr1.

The board is routed in the component (top) layer with a few jumpers in the solder (bottom) layer. The component side and solder side of the

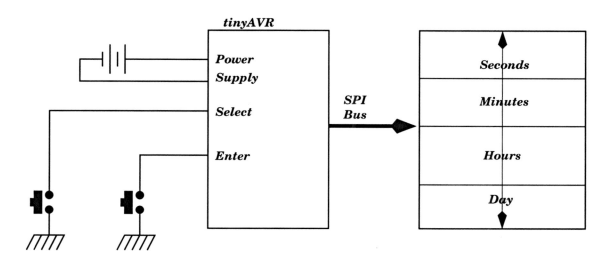

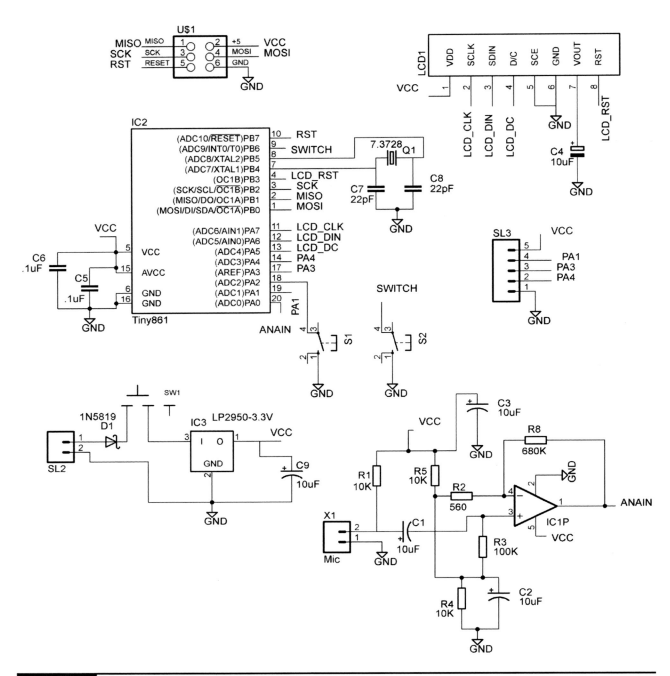

Figure 4-16 Zany clock: Schematic diagram

board are shown in Figures 4-17 and 4-18, respectively. Two working demonstrations are shown in Figures 4-19 and 4-20.

Design Code

The compiled source code, along with the MAKEFILE, can be downloaded from www.avrgenius.com/tinyavr1.

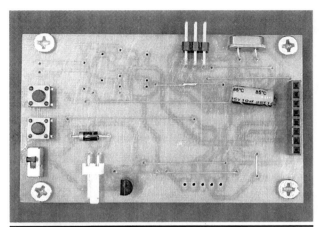

Figure 4-17 Clock PCB: Component side

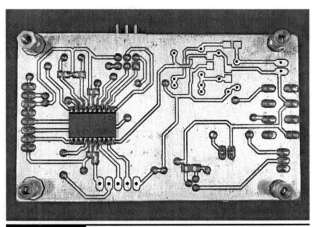

Figure 4-18 Clock PCB: Solder side

The code runs at a clock frequency of 7.3728 MHz. The controller is programmed using STK500 in ISP programming mode. To run the device on an external clock, CKSEL fuses must be programmed to "1101" before programming the controller. Certain parts of the code are common to previous projects involving the use of the NOKIA 3310; thus, we have included the LCD library in our project, thereby eliminating the need to write the LCD interfacing code all over again. The functions related to interfacing the LCD and displaying data on the LCD are taken directly from this library.

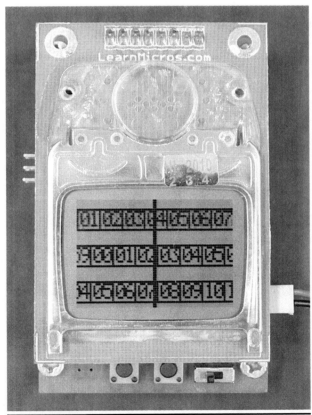

Figure 4-19 Project with display showing three parameters

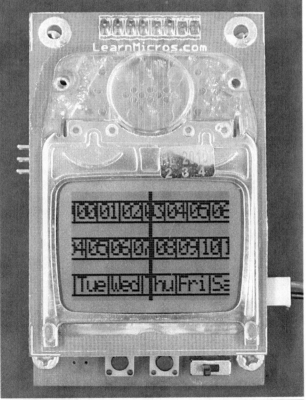

Figure 4-20 Project with display with scrolled values

Now to generate the display pattern as a timeline, we have created two functions: **boxes** and **centerline**. The former draws the outer boundary of time quantities on the zero, second, and fourth page, whereas the latter draws a centerline used to view the elapsing time. The centerline has X coordinates of 41 and 42, and is present on every page. Data related to seconds, minutes, and hours is stored in a constant array table in program memory. A particular character or digit requires 12 columns for display, and the table consists of 60 such characters, thus requiring a total of 720 columns. Similarly, table2 contains data related to days. The display block of each day requires 18 columns; thus, the total size of the array is $18 \times 7 = 126$.

TIMER0 is initialized with a frequency of 7200 Hz, obtained by prescaling the system frequency by 1,024, and the overflow interrupt is enabled by setting **TOIE0** bit in **TIMSK** register. Further prescaling is achieved with the software as shown in the code segment of the **ISR**:

```
ISR(TIMER0_OVF_vect)
    //timer interrupt service routine
{
    TCNT0L=(255-225);
    count++;
    if(count==8)
        //software prescaling by 8
    {
        count=0;
        if(s.c==0)//sec count
        s.c=1;
            if(m.c<20)//min count
            m.c++;
            if(h.c<1200)//hour count
            h.c++;
            if(d.c<19200)//day count
            d.c++;
    }
}
```

In this routine, the timer register is initialized such that the timer counts 225 and not 255 when it overflows. The timer runs at a frequency of 7,200

Hz; thus, an overflow interrupt occurs at 225/7,200 sec. Now the main content of the **ISR** is executed when **count** reaches a value of 8. This means that the content is executed every $225 \times 8/7,200$ of a second, which is nothing but one-quarter of a second.

A particular second elapses when its 12 columns surpass the timeline or centerline. Thus, on every eighth timer interrupt, that is, on one-quarter of a second, three columns should surpass the timeline. This is done in the main loop of the program. An infinite while loop keeps the track of seconds to be displayed on the LCD. In this infinite loop there are four functions from the **rtc library**. These are seconds(), minutes(), hours(), and days(). Out of these, seconds() is explained next:

```
if(s.end!=723)
    //if end is not equal to the end
    //limit of table1
{
cursorxy(0,s.row);
    //put cursor on the seconds page
for(i=s.start;i<s.end;i++)
    //write contents from start to end
{
column=pgm_read_byte((table1+i));
writedata(column);
}
centerline();       //display centerline
if(s.c==1)          //check count
{
s.start+=3;
s.end+=3;
s.c=0;
}
}
else if(s.end==723)
    //if end reaches the limit of table1
{
centerline();
cursorxy(0,s.row);
for(i=s.start;i<(s.end-3);i++)
    //display contents from start
{
column=pgm_read_byte((table1+i));
writedata(column);
```

```
}
for(i=0;i<s.x;i++)
    //display from the first element
    //of table 1
{
column=pgm_read_byte((table1+i));
writedata(column);
}
centerline(); //timeline
if(s.c==1)      //check count
{
s.start+=3;
s.x+=3;
s.c=0;
}
if(s.x==84)
    //check if start has reached end
    //or not
{
s.end=84;
s.start=0;
s.x=3;
}
}
```

Every display quantity, be it seconds, minutes, hours, or days, has its set of variables defined in **rtcpara** structure. These variables are defined as:

- **start** Starting X-coordinate of the data to be displayed.

- **end** End value of the X-coordinate of the data to be displayed.

- **x** X offset used to display data from the beginning of the PROGMEM table when end reaches its limiting value.

- **c** Count of the corresponding quantity. This count is incremented in the **ISR** noted earlier.

- **row** Represents the page or the bank of the corresponding quantity.

In the previous routine, the first part is executed when **s.end** is not equal to the limiting value, that is, 723. Then data is printed on **s.row** page from **s.start** to **s.end**, which is 84 columns apart; hence, data is printed on the full page. Then **centerline** or

timeline is flashed to view the elapsing time, and the **s.c** is checked. If **s.c** is found equal to 1, **s.start** and **s.end** are incremented by 3 and **s.c** is made 0 again. The second part is executed when **s.end** reaches its limiting value. At first, data is printed from **s.start** to **s.end – 3** (because **s.end** was incremented by 3 in the first part) on the **s.row** page. Then data is printed from the start of the PROGMEM table to **s.x**. This continues until **s.x** reaches a value of 84. At this point, **s.start** is made equal to **0** and **s.end** is made equal to **84**. Similar checking on **s.c** is performed as in the first part, and **centerline** is flashed.

Minutes, hours, and days are manipulated and displayed using similar routines defined in **rtc.c**. Apart from this, the code consists of initial time setting through hardware switches and using the pin change interrupt to toggle between displaying days or seconds.

Working

To use the clock, simply power up and set the time. Then you see the seconds fly by on the screen. The minutes scroll slower compared to the seconds, and the hours are even slower. Press S2 to see the day of the week. Press S2 again to get back to the original display.

Project 19
Rise and Shine Bell

This is a project we designed for a residential school. They wanted a really loud alarm bell to wake the kids up for morning exercises. We added more features by providing additional alarm functions. A separate code programmed in the same hardware converts the alarm bell into a school bell. The next illustration shows the block diagram of the alarm bell. It has a mains power supply input as well as a battery backup. The audio amplifier provides a loud output to wake the kids up from sleep.

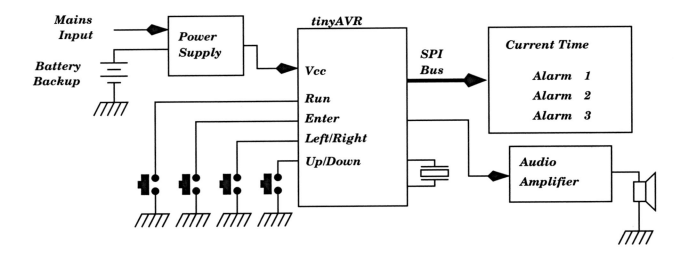

Design Specifications

The objective of the project was to design a loud mains-powered alarm bell with battery backup for the timekeeping functions. The bell offers three alarm settings.

Design Description

Figure 4-21 shows the main schematic of the project, and Figure 4-22 shows the schematic diagram of the add-on switches.

The alarm bell has two power supply inputs: a source from mains power and another from a battery. The mains-powered source is used to power the microcontroller circuit as well as the audio amplifier, while the battery source only powers the microcontroller. The microcontroller, a Tiny861, uses a 32.768-KHz crystal as the system clock source to execute the program as well as to maintain time. Providing operating voltage to the microcontroller is, therefore, extremely important and critical for the operation of the alarm bell. The user can set up to three alarms, and when any of the alarm times matches the current time, the microcontroller generates a tone on the PB3 pin that drives a 20W audio amplifier. The microcontroller power supply is from a LP2950-3.3 regulator since the system interfaces

to a Nokia display. The display is used to interact with the user.

On powering up the circuit, the current time set appears on the top-right corner of the LCD along with the menu, which has six items—TIME, MODE, DISP, ALARM1, ALARM2, and ALARM3.

- **TIME** Configures the present time of the system.

- **MODE** Switches individual alarms on and off.

- **DISP** Switches off the display.

- **ALARM***x* Sets the time of the individual alarm.

The triangle-shaped pointer shows the current menu item selected, and it can be moved down by pressing S2 and up by pressing S3. Pressing S1 displays the submenu of the selected menu item. At this level, S4 has no function but to asynchronously stop the running alarm. The submenus are as follows:

- **TIME** S1 updates the current time shown and exits the menu. S2 changes the pointer of the digit to be changed. S3 and S4 increment and decrement, respectively, the digit currently pointed to.

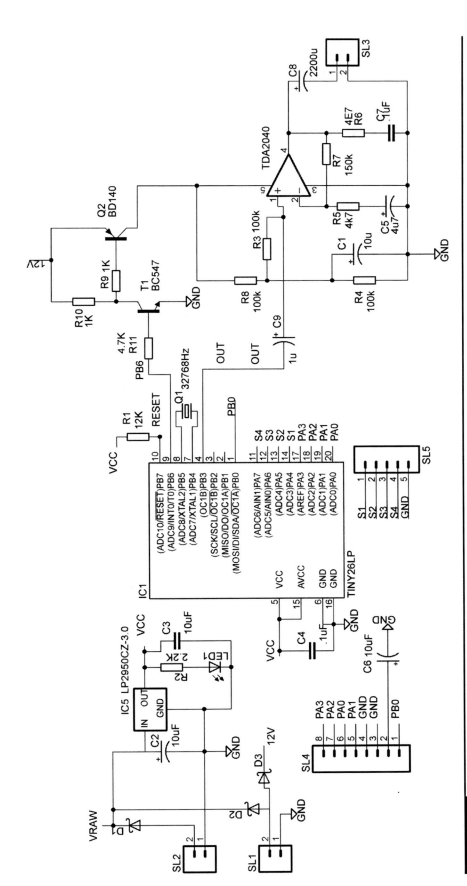

Figure 4-21 Rise and shine bell: Schematic diagram

125

LISTING 5-1   AVR microcontroller's assembly-language firmware *(continued)*

```
                    brne dec_r20
                    dec r19
                    brne dec_r20
                    ret
min_delay: in r0, SREG
                    ldi r18, 200
not_over:
                    dec r18
                    brne not_over
                    out SREG, r0
                    ret
```

LED is forward-biased (to light up the LED) for a time equal to the time stored in register R19. Thus, if the microcontroller measures T time units as the time it takes to discharge the LED in the first measurement cycle, it turns the LED on for T time units. The frequency at which the LED is pulsed is proportional to the light incident on the LED.

Working

The circuit was tested by applying light of known intensity through a test LED. For low values of LED forward current, the light output intensity is fairly linear. The light output of the test LED was coupled to the sensor LED (LED1 in Figure 5-3) of the circuit. It was ensured that no other external light was incident on the sensor LED by enclosing the test LED and the sensor LED in a sealed tube covered with black tape. The test LED current varied between 0.33mA and 2.8mA. The corresponding output of the sensor LED flashing frequency was recorded and is shown as a plot in Figure 5-4. As can be seen in this figure, the circuit provides a fairly linear output.

The ATTiny15 AVR microcontroller is an eight-pin device. The circuit presented here uses only three out of the six I/O pins. The rest of the pins can be used to control external devices or for communication with other devices. The efficiency of using an LED as a sensor depends upon current

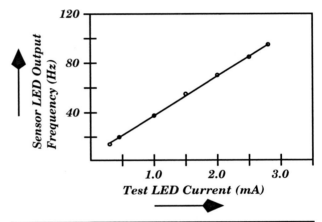

Figure 5-4   Plot of LED sensor output as a function of ambient light intensity

source and capacitance values of the LED operated in reverse-bias. We estimated these values to compare with the figures reported in literature. To estimate the reverse photocurrent, we connected a 1-meg-ohm resistor in parallel with a sensor LED and measured the voltage across the resistor. The sensor LED was subjected to constant illumination and voltage across the resistor noted. We changed the resistor value to 500 kilo-ohm and 100 kilo-ohm and repeated the measurement. The resultant photocurrent for the constant illumination was observed to be around 25mA for all the measurements. For the same illumination on the sensor LED, the frequency generated by the circuit in Figure 5-3 was measured, and delay loop times, current, and voltage were substituted in the equation dv/dt = I/C to calculate the reverse capacitance. The calculated values lie in the range of 25 to 60pF.

Project 21
Valentine's Heart LED Display with Proximity Sensor

The use of the LED as a sensor is further illustrated in this project with a captivating output that will mesmerize the viewer. The project consists of several LEDs arranged in a heart formation, "throbbing" at a normal rate depicting the throbbing of a human heart. However, if a hand is brought close to the blinking LEDs, the sensor LED detects the reduction in the incident light and the microcontroller increases the throbbing rate. If the hand is brought closer, almost touching the LED matrix, the sensor LED detects this and the microcontroller flashes the LEDs in an unusual fashion to indicate a "happy" mood. Once the hand is taken away, the "heart" resumes normal "throbbing." The illustration shows the block diagram of the LED heart display.

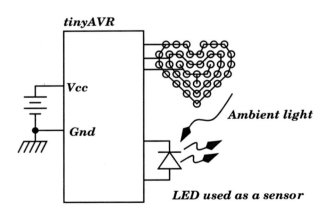

Design Specifications

The objective was to create an interactive project in the form of an LED matrix arranged in a heart shape. The interactivity is based on sensing the proximity of a hand or any other object, and is used to change the blinking rate of the LEDs. The

project is battery operated so as to make it portable.

Design Description

The schematic diagram of the project is shown in Figure 5-5. The circuit consists of 27 LEDs arranged in three concentric layers. The outermost layer consists of 14 LEDs. The middle layer consists of ten LEDs, and the innermost layer consists of three LEDs. At the center of the inner layer is another LED, marked LED28. This LED is not being used to emit light but to sense the ambient light. The circuit is battery powered and has an on-board 3.3V voltage regulator. Each LED has a 560-Ohm resistor in series. Thus, the current through each LED is about 3mA. Three NPN transistors (BD139) are used to switch each of the three layers of the LEDs on or off. The BD139 NPN transistor is capable of handling 1A of collector current. However, in the circuit, the maximum current flows only in the outer layer of LEDs and is about 45mA, so any NPN transistor with 100mA collector current rating would also work fine.

The ambient light intensity is sensed by LED28, which is connected between two microcontroller pins (pin PB3 and PB4). The microcontroller uses this LED to sense the ambient light by reverse-biasing the LED and measuring the time it takes for the logic "1" to discharge to logic "0." The circuit can be powered with an external battery with voltage between 5 and 10V.

Fabrication

The board layout in EAGLE, along with the schematic, can be downloaded from www.avrgenius.com/tinyavr1.

The project has been fabricated using a custom, single-sided circuit board. The photographs of the completed project (component side and solder side) are shown in the next illustrations. During

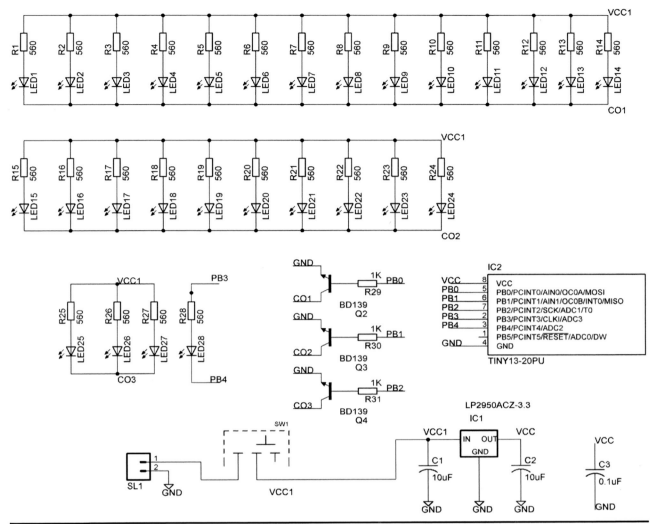

Figure 5-5 LED heart display: Schematic diagram

soldering, all the SMD components (resistors, etc.) were soldered first, followed by a few jumper wires. Subsequently, all the LEDs, transistors, a socket for the microcontroller, and other components were soldered.

Design Code

The compiled source code, along with the MAKEFILE, can be downloaded from www.avrgenius.com/tinyavr1.

The code runs at a clock frequency of 9.6 MHz. The controller is programmed using STK500 in ISP programming mode. The LED heart works in one of three possible modes: NONE, NORMAL, and HAPPY. These modes are differentiated by the blinking patterns of the LEDs. The main infinite loop of the program is given here. The value of the variable **a** is set to 20 globally.

```
while(1)
    {

        if(mode==NORMAL)
        {
            PORTB &=0b11111000;
                //off
            mydelay(a+a+a+a);
            PORTB|=0b00000111;
                //on
            mydelay(a+a);
            PORTB &=0b11111000;
                //off
            mydelay(a+a);
            PORTB |=0b00000111;
                //on
            mydelay(a+a);
            time=0;

        }
        else if(mode==HAPPY)
        {


            if(state==1)
            PORTB |=1<<2;
            else if(state==5)
```

```
            PORTB &= ~(1<<2);

            if(state==5)
            PORTB &=~(1<<1);
                //Middle off
            else if(state==2)
            PORTB |=1<<1;
                //Middle on


            if(state==4)
            PORTB &=~(1<<0);
                //Outer off
            else if(state==3)
            PORTB |=1<<0;
                //Outer on


            mydelay(5*state);

            state++;
            if(state==6)
            state=1;

            time++;
            if(time==100)
            {
                mode = NORMAL;
                previous_mode
                = NONE;
                time=0;
                a=20;

            }
        }
        check();
    }
```

If the mode is set to NORMAL, the system starts toggling between the ON and OFF states of the LEDs, with some delay depending on the value of **a**. The other case is when the mode is set to HAPPY. There are three parts to the LED heart: outer heart, middle heart, and inner heart. When the mode is set to HAPPY, depending on the value of the variable state, the LEDs in one of the three parts of the heart are either turned on or off.

The value of the variable **mode** is set to NONE globally. The system calls a function **check()** for proximity sensing, which modifies the value of the mode accordingly. This function is shown next. The value of the count from TIMER0 is stored in the variable **i** until there is a logic level 1 at the third bit of PINB (which is connected to the sensing LED), so as to detect the proximity of any opaque object (like a hand, in this case) that might be blocking the light falling on the LEDs. Depending on the distance of the object from the LEDs (which can be determined from the count value) and on the previous mode in which the system was working, the current mode is either set to NORMAL or HAPPY.

```c
void check(void)
{


        TCCR0B = 0;//stop timer
        TCNT0 = 0;//clear timer
        i=0;
        DDRB = 0b00010000;
        PORTB &= ~(1<<3);
        i=0;
        TCCR0B = 1<<CS02|1<<CS00;
            //Prescaled by 1024
        while((PINB&(1<<3))&&(i<80))
        {
                i=TCNT0;
        }

        if(i>50)
        {

    if((previous_mode==NORMAL))
            {
                    mode = HAPPY;
            }
            else a=20;
        }
        else if(i<50)
        {
                if(i<40)

                    previous_mode = NORMAL;
            }
```

```c
            else previous_mode = NONE;
            mode = NORMAL;
            a=i/3;

        }


        PORTB |=1<<3;
        DDRB = 0b00011111;


}
```

The other parts of the code involve, as usual, the initializations for the various parameters. The **mydelay** function is defined here; if the system is working at a frequency of 9.6 MHz, it provides a delay of 10 ms for an argument of 1.

```c
void mydelay(int var)
    //delay of 10ms for var=1 at
    //9.6 MHZ
{

        unsigned char il, jl, kl;

        for (il=0; il<var; il++)
        for (jl=0; jl<251;jl++)
        for (kl=0; kl<50; kl++)
        asm("NOP");

}
```

Working

To use the touch-sensitive throbbing heart, just hold the circuit board by your chest (put the batteries in the adjoining pocket) and turn it on. The LEDs should start blinking at a normal rate. Now bring a hand close to the blinking LEDs, and the throbbing rate should start increasing. Keep bringing the hand closer to the circuit board and eventually touch the board and then take your hand away. Once you take your hand away, the LEDs should start blinking in the HAPPY mode, returning to normal throbbing after a few seconds.

Project 22
Electronic Fire-free Matchstick

Imagine a matchstick that you strike across a matchbox and it starts flickering but without any fire? Well, what good could be such a matchstick? How about for use in plays and theaters and for kids who you don't want to play with fire? The electronic fire-free matchstick is such a device, in that, you have to strike the matchstick across a matchbox and then the matchstick starts glowing. The project utilizes an inductor on the matchstick and a hidden magnet inside a matchbox to achieve this feat. The following illustration shows the block diagram of the matchstick.

Design Specifications

The objective was to create a chargeable and portable "matchstick" that one could strike across a matchbox to light up. Like a normal matchstick, the electronic matchstick is supposed to light up only for a short duration, extinguishing after this time. There are many application areas for such a seemingly useless project, especially to do with kids and in theater.

Design Description

Given the requirement that a light is to start glowing when the matchstick is struck against a matchbox, two questions arise: how to power the matchstick and how to generate the trigger that lights up the matchstick. There are several ways to power the matchstick. Batteries seem a natural choice. However, using batteries would entail the use of an on/off switch, but more importantly, how would you turn the matchstick off? The microcontroller program could be designed to turn off the light after a preset amount of time, but that would take the randomness out of the matchstick glow. So, instead of a battery, we decided to use a supercapacitor. A supercapacitor can store a lot of energy that can be used to provide suitable voltage and current with the help of a DC-DC converter. The advantage of using a supercapacitor is that it would discharge naturally as its energy is utilized and drained through the circuit and thus the matchstick would shut off on its own.

The second aspect is how to trigger the matchstick to start working. For that, a simple inductor-based trick is used. As we have seen earlier, moving an inductor in a magnetic field produces a voltage across the inductor. If the rate

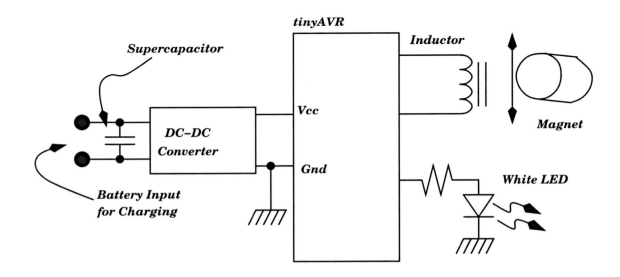

change of the magnetic field is large enough, it can produce a sufficiently large voltage to interrupt the microcontroller! Once interrupted, the microcontroller can be made to do anything we want.

Figure 5-6 shows the schematic diagram of the matchstick. A 10F supercapacitor is used to power the circuit. Its voltage rating is 2.7V, which means it can be charged to 2.7V, storing a total of 27 coulombs of charge. To charge the supercapacitor, an external battery is used (2 × 1.5V alkaline or NiMH batteries in series are most suitable). The supercapacitor drives a boost-type DC-DC converter, the MAX756, which is configured to provide 5V output voltage. The MAX756 can be set to provide 3.3V output voltage also, but since we want to drive a white LED, a 3.3V supply voltage would not be suitable for that. The DC-DC converter uses an inductor L1 (22μH) and D1 Schottky diode (1N5819) for its operation. Once the supercapacitor is charged above 1.2V, the DC-DC converter provides 5V output voltage to the Tiny13 microcontroller. The microcontroller is powered and is waiting for an external trigger. The trigger is received on the second inductor in the circuit L2, which is an inductor with a large value (compared to L1) at 150mH. When the matchstick is struck across the face of a matchbox with a hidden magnet, it produces a voltage spike across the inductor. The diode across the inductor ensures that any negative voltage generated across the inductor doesn't harm the microcontroller. Once the microcontroller is triggered, it executes a code that lights up the LED in a random fashion (just like in the LED candle project). Lighting the LED consumes a much larger amount of energy than just powering the microcontroller, and this discharges the supercapacitor rapidly. Once the supercapacitor voltage drops below 0.7V, the DC-DC converter is unable to provide +5V supply voltage, and thus the microcontroller stops working and the matchstick "extinguishes." A 10-mm white LED is used in the project, although a smaller 5-mm LED can also be used.

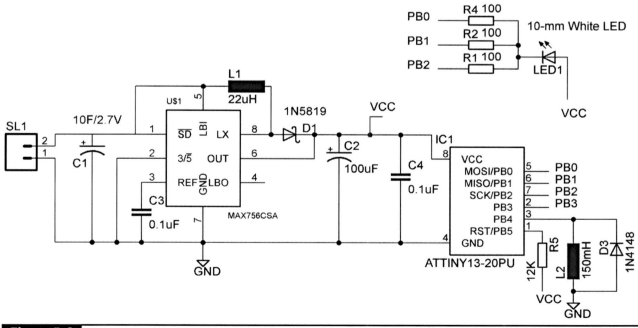

Figure 5-6 Electronic fire-free matchstick: Schematic diagram

Fabrication

The board layout in EAGLE, along with the schematic, can be downloaded from www.avrgenius.com/tinyavr1.

The circuit is fabricated using a custom circuit board, keeping in mind the unique requirement of the project. The circuit board is designed so that it is narrow and long so as to appear as a matchstick. It is conveniently packaged in a transparent Perspex tube, as seen in the next illustration. The PCB is soldered with all the SMD components first, followed by other leaded components. Once soldered, it is put inside the tube. The tube length is chosen such that the LED protrudes from the other end. The LED end of the tube is sealed with hot glue melt and may be painted red.

Design Code

The compiled source code, along with the MAKEFILE, can be downloaded from www.avrgenius.com/tinyavr1.

The code runs at a clock frequency of 1.6 MHz. The main infinite loop of the code is shown next. If the variable mode is set to "ON," the system generates a pseudo-random variable called **lfsr**

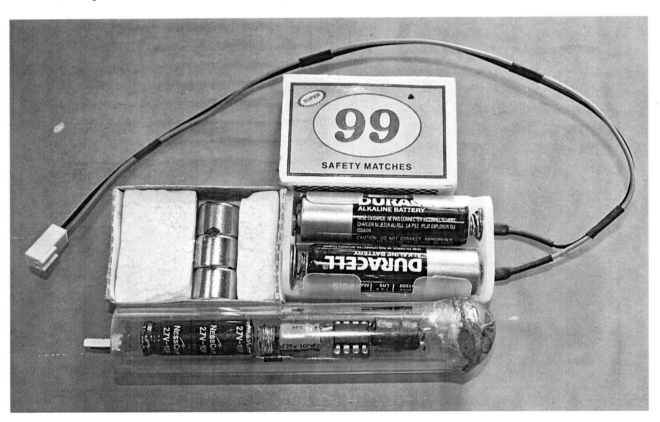

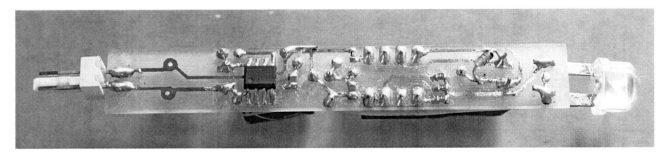

using a linear feedback shift register (LFSR) of 32 bits with taps at the 32nd, 31st, 29th, and 1st bits. This value is stored in the variable temp (so as to maintain the last state of the LFSR), and the value of temp is sent to the output at PORTB. The delay introduced into the system is also a function of the variable temp, and hence, is also correspondingly pseudo-random.

```
while(1)
{

i=1;//This is made to ensure that
    //interrupts before this are
    //neglected
        if(mode==ON)
        {
            //Galois
            lfsr = (lfsr >> 1) ^
            (-(lfsr & 1u) &
            0xd0000001u);
            /* taps 32 31 29 1 */
            temp = (unsigned char)
            lfsr;
            DDRB= ~temp;
            PORTB = temp;

            temp = (unsigned char)
            (lfsr >> 24);
            _delay_loop_2(temp<<7);
        }


}
```

The value of the variable mode is, however, globally set to "OFF." The main program sets the variable **i** to 1. When the matchstick is struck across the matchbox, it produces a voltage spike across the inductor, which interrupts the processor and the PCINT0 ISR is executed. In the ISR code, the value of mode is set to "ON" and the masks GIMSK and PCMSK set to 0✕00, using the interrupt service routine shown next. Once the program returns to the main code, the infinite loop executes the Galois LFSR code, which lights up the LED in a random fashion.

```
ISR(PCINT0_vect)
{
    if(i==1)
    {
        mode = ON;
        GIMSK = 0x00;
        PCMSK = 0x00;
    }
}
```

The other parts of the code are the various initializations, which provide values for the masks and variables used in the program.

Working

To use the matchstick, one must have the special matchbox with the hidden magnet. The polarity of the magnet is important, that is, which pole of the magnet faces outwards. The supercapacitor in the matchstick has to be charged before the matchstick can be used. For that, we use a battery holder to connect two AA batteries in series. The battery voltage is used to charge the supercapacitor. After the batteries are connected to the supercapacitor, it may take some time for the supercapacitor to charge fully. Once the capacitor is charged (you may confirm that by measuring the voltage across the supercapacitor, which must be more than 2V for acceptable operation of the matchstick), you are ready to strike the matchstick across the matchbox. You will notice that it is not necessary for the matchstick to be physically rubbed across the matchbox. As long as you rapidly strike the matchstick in the vicinity of the matchbox, it will produce a voltage spike across the inductor to trigger the operation. Please see the video at www.avrgenius.com/tinyavr1 for proper operation in case you cannot get your matchstick to operate as expected.

Project 23
Spinning LED Top with Message Display

Several LED spinning tops are available. They usually have LEDs of different colors on them, and as you spin them, these colors spread across the circumference of the spinning top. However, our top is totally different from these other available tops. Our top displays a message when you spin it. Better still, when you spin the top in the opposite direction, it can display a different message. The LED lighting pattern is changed rapidly to display the required message and the persistence of vision of the human eye is able to record the displayed message. The original design was published as an article in *Elektor* magazine in December 2008 ("LED Top with Special Effects"). We adapted that circuit and made some improvements such that our spinning top can display a message when you spin it one way and a different message when you spin it the other way. The *Elektor* top could only

display the same message if you spun the top one way. The illustration shows the block diagram of the spinning top. It consists of a row of SMD (surface mount device) LEDs arranged radially from the center of the top towards the periphery. The top is powered with two AAA NiMH or alkaline batteries providing between 2.4 and 3V. Since the microcontroller is to be powered with 5V, there is an on-board step-up type DC-DC converter, which boosts the battery voltage to +5V. To determine the motion and to determine the direction of motion of the top, the circuit employs motion-sensing components using a couple of inductors.

Design Specifications

The objective of the project is to create a spinning top with LEDs that will display messages when the top is spun. The top is to be battery powered and should display different messages when it is spun in the two directions. The circuit uses eight LEDs to display text as well as graphics information.

Design Description

Figure 5-7 shows the schematic diagram of the direction-aware LED spinning top. It is powered with two AAA batteries labeled AAA1 and AAA2 in the schematic. An on/off switch, SW2, allows the power to be switched off when not in use. Beware that there is no power-on indicator and it is likely that you will forget to switch the power off and thus drain the batteries.

The battery voltage is applied to the MAX756 DC-DC converter, which provides +5V supply voltage to the rest of the components on board using inductor L1 and Schottky diode D1. The supply voltage of +5V is used to power the microcontroller AVR ATTiny44 (IC3) as well as two dual op-amps LM358 (IC1 and IC2). The microcontroller drives eight SMD LEDs in current sink mode. The motion detector circuit consisting

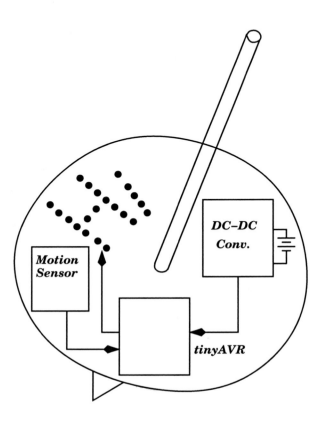

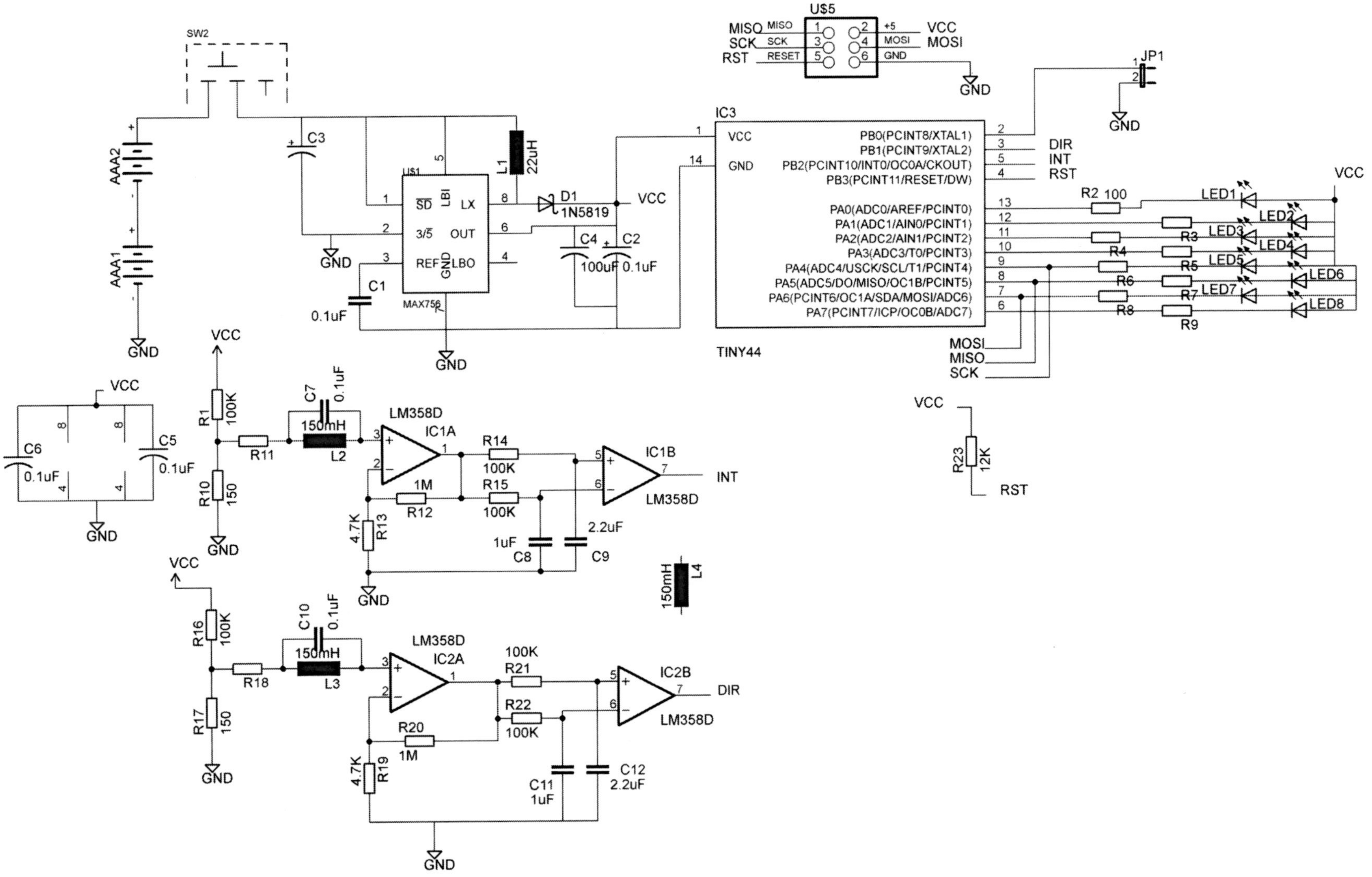

Figure 5-7 Direction-aware LED spinning top: Schematic diagram

of the two op-amp ICs indicates whether the top is spinning or not and, if it is spinning, the direction of the spin. This information is provided to the microcontroller on signal pins INT (pin PB2) and DIR (pin PB1). The motion and direction detector circuit consists of two identical channels comprising inductors L2 and L3. When an inductor moves in a magnetic field, it produces a voltage. The circuit exploits the magnetic field of the Earth. The inductors are arranged such that they are perpendicular to the Earth's magnetic field. Thus, the inductors produce a tiny AC voltage when the top spins. The frequency of this AC voltage is equal to the rate at which the top spins. The two inductors are placed 90 degrees apart on the circumference of the edge of the top. Thus, the waveform of one inductor is 90 degrees out of phase compared to the waveform of the other inductor. If the sinusoidal waveforms produced by each of the inductors is observed on an oscilloscope with a common time base and one of the inductor waveforms is taken as a reference waveform, then the other waveform will lead the reference waveform in one direction of motion and lag the reference waveform in the other direction, as shown in the next illustration.

The sinusoidal signals from the inductors are amplified by the op-amps configured as noninverting amplifiers (IC1A and IC2A, respectively). These amplified outputs are converted into a square wave by comparing each waveform with a delayed version of the same waveform using RC delay circuits (R14-C9 and R15-C8 for one channel and R21-C12 and R22-C11 for the other channel). These delayed waveforms are passed through comparators (IC1B and IC2B, respectively) to get square waveforms of the same frequency as the sinusoidal signals from each of the inductors. The rectangular signals from the op-amps (INT and DIR) are connected to the microcontroller. The INT signal is connected to the PB2 pin of the Tiny44 microcontroller, which is an interrupt input. The pin is configured as a

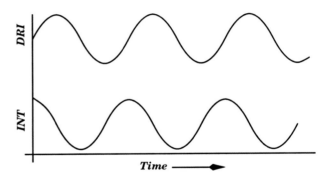

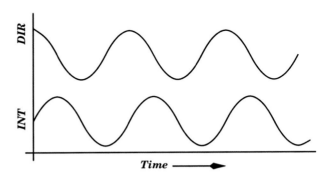

rising-edge input by the microcontroller code. The other channel DIR is connected to PB1. Each time the microcontroller is interrupted by the INT signal, it executes an ISR, and in the ISR it reads the state of the DIR pin. If DIR pin is "0," then it assumes one direction of motion and if DIR is "1," then it assumes the opposite direction. The microcontroller also measures the frequency of the interrupting signal INT. It uses this information to get an idea of the time it takes for the top to complete one rotation. The LED lighting patterns for each direction are stored in the program memory of the microcontroller. The microcontroller uses the spinning rate information to decide how long a particular LED pattern should persist. If the top speed reduces, the LED pattern time is proportionately increased. If the top speed is high (as will be the case at the beginning of the rotation), each LED lighting pattern is set for a smaller duration. This ensures that the displayed message remains uniform irrespective of the top speed of rotation.

The six-pin connector is the ISP connector that is used to program the microcontroller. Jumper JP1 is put aside for future use, and the current version of the program does not read the state of the PB0 pin connected to the jumper. Inductor L4 is simply to balance the top. It is electrically disconnected from the circuit.

Fabrication

The board layout in EAGLE, along with the schematic, can be downloaded from www.avrgenius.com/tinyavr1.

The top is built using a custom printed circuit board. The next illustrations show the top and bottom sides of the circuit board, respectively.

All the inductors are placed on the bottom side of the circuit board and are covered with hot glue melt to prevent damage. Inductors L2 and L3 are wound with 42 SWG of copper enameled wire. This is a very thin wire, and extreme care is to be exercised when winding the coil. The dumbbell-shaped ferrite former was completely filled with copper wire, and the resultant inductance was about 150 mH. Details of the dumbbell (10 mm

height and 3 mm inner diameter) and how to wind copper wire on it are available on our website. L1 is wound with 28 SWG with 20 turns to get about 22uH of inductance. The number of turns on L4 is not critical as long as it has the same weight as the other inductors. All the inductors are placed on the periphery of the circular PCB, 90 degrees apart. A working graphic of the spinning top is shown here:

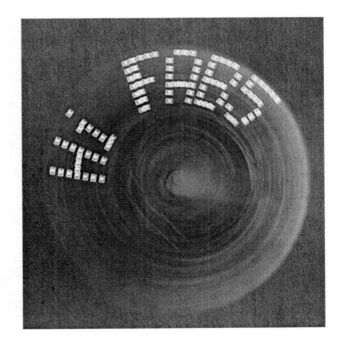

All the SMD components are soldered first, followed by the leaded components. Then the battery connector pins are soldered, and finally the inductors are soldered on the bottom side. A solid plastic rod, machined to fit in the center hole of the circuit board, forms the axle of the top.

Design Code

The compiled source code, along with the MAKEFILE, can be downloaded from www.avrgenius.com/tinyavr1.

The code runs at a clock frequency of 1 MHz. The maximum and minimum times one rotation might take are predefined, as are the strings to be printed on clockwise spinning as well as counterclockwise spinning. The main function initializes the interrupts and enters the infinite loop, from where it calls the **double_string_display** function repeatedly. This function is shown below.

This function uses the **set_leds** function to display the required pattern on the LEDs according to which direction the top is spinning

```c
void double_string_display (void)

{

        // Initialize display
        construct_display_field();

        // TOP runs
        while (running_condition() == TOP_TURNING)
        {
                i = current_column/2;
                if(current_column%2==1)
                {
                        set_leds(LED_ALL_OFF);
                }
                else if(current_column%2==0)
                {
                        if(mode==CLOCKWISE)
                        {
                                if(i<=(STRING_LENGTH1*6))
                                set_leds((display_field_clock[i])&0x7F);
                                else
                                set_leds(LED_ALL_OFF);

                        }
                        else if(mode==ANTICLOCKWISE)
                        {
                                if(((STRING_LENGTH2*6)-i)>=0)
                                set_leds((display_field_clock[(STRING_LENGTH2*6)-i])&0x7F);
                                else
                                set_leds(LED_ALL_OFF);
                        }

                }
        }

        // TOP does not run
        while(running_condition() != TOP_TURNING)
        {
                // Disable all leds
                set_leds(LED_ALL_OFF);
        }
} /* double_string_display */
```

in, and which column is to be lit up. The **construct_display_function** fills the array **display_field_clock** according to the text and direction. The timer and clock values for the proper display of the strings (so as to be detected by the persistence of vision), as well as the mode the top is currently spinning in (clockwise or counterclockwise), are modified in the INT0, TIMER0 overflow, and TIMER1 overflow interrupts. Every alternate column display is left blank to ensure that the displayed columns do not merge with each other due to the high speed of rotation of the top. The ratio of columns left blank to columns displayed determines the width of the displayed characters.

Working

Once the top is soldered, the batteries are installed and the microcontroller is programmed with the application code available from our website, using the ISP interface. Once the microcontroller is programmed, the ISP cable is removed and the power switch is turned on. Gently hold the axle of the top between the palms, spin the top, and let it go on a solid flat surface. You will see messages appear. It may take a bit of practice to spin the top. Let the top stop spinning, and now spin it in the other direction and a different message will appear. Make sure you turn the power switch off when the top is not in use.

Project 24
Contactless Tachometer

A tachometer used to measure the rotational speed of a motor often requires physical contact with the motor shaft. Sometimes, however, it is desirable to measure the rotational speed of the motor without any physical contact with the motor or its shaft. In some noncontact tachometer schemes, a mirror attached to the rotating part of the motor is used to reflect a pulsed beam of laser light, which is then measured with an optoelectronic circuit. Other contactless methods of measurement involve measuring the induced voltage over the sparkplug wire of the motor or the engine, if available.

We present a contactless tachometer circuit here that does not use any pulsed laser light or any such method. Instead, our method involves attaching a small magnet to the shaft or any other rotating part of the motor and measuring the period of the voltage generated across a stationary inductor in the field of the rotating magnet. With this method, the measurement circuit is electrically as well as physically disconnected from the motor or its shaft, or any other part of the motor. The block diagram of this project is shown here:

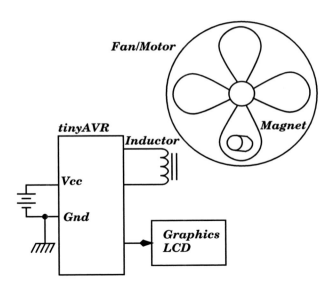

Design Specifications

The objective of the project is to design a tachometer that could be used to measure the rotational frequency of fans, motors, etc., without any mechanical or electrical contact with the object. The system is to be battery operated with a graphics display to display the necessary information about rotational speed, period, etc.

Design Description

The circuit consists of a large inductor (100mH) followed by a large-gain op-amp configuration, similar to the circuit in the spinning top project. The output of the op-amp is passed through two low-pass filters with different time constants so as to generate a small phase difference between the two filtered output signals. These two phase-differing signals are fed to another op-amp in a comparator configuration to convert the original signal to a pulse output signal. Now the signal that the inductor provides depends on any varying magnetic field in the vicinity. If the circuit is placed close to a magnet mounted on a shaft of the motor, then as the motor moves, the magnetic field perceived by the inductor would vary and this would induce a sinusoidal voltage across the inductor.

Figure 5-8 shows the circuit diagram of the contactless tachometer. The induced voltage across the inductor is amplified using one section of op-amp IC3-A (LM358), and the output is passed through two RC filters (arranged in parallel) with different time constants. The output of these filters is applied to the two inputs of the second section (IC3-B) of the LM358 op-amp. This converts the incoming sinusoidal signal from the inductor into a pulse waveform of the same frequency. The pulse waveform is applied to the AVR Tiny45 microcontroller, which is set up to measure the frequency of the signal applied to one of its pins (pin 3) configured as input. The other pins of the microcontroller control a dot matrix LCD display from the hugely popular Nokia 3310 mobile phone discussed in a previous chapter. The tachometer is powered with a 9V battery connected to the SL3 connector.

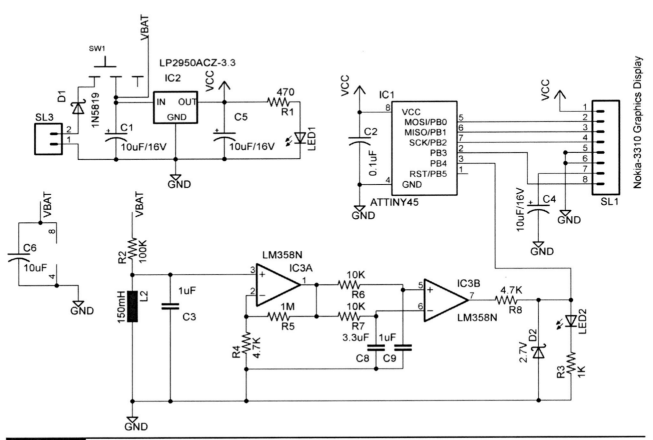

Figure 5-8   Contactless tachometer: Circuit diagram

The output of the display is shown in the illustration on the next page, which also shows the completed circuit in a suitable enclosure. The monochrome display does not have any backlight, so external white LEDs are arranged on the two sides of the display to allow visibility in low light conditions.

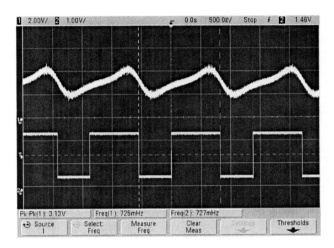

The circuit is powered with a 9V battery connected to the SL3 connector, and this voltage is also used to power the LM358 op-amp. The Nokia display requires a DC power supply voltage between 3 and 3.3V, and so an LDO LM2950-3.3V is used to derive the supply voltage for the display as well as the microcontroller. The output of the op-amp is passed through a resistor and a 3V Zener diode to limit the signal applied to the microcontroller.

Fabrication

The board layout in EAGLE, along with the schematic, can be downloaded from www.avrgenius.com/tinyavr1.

The component and solder sides of the soldered board are shown in the next two illustrations, respectively. The final illustration shows the completed assembly of the project.

Design Code

The compiled source code, along with the MAKEFILE, can be downloaded from www.avrgenius.com/tinyavr1.

The maximum time one rotation might take is predefined. The function **running_condition** checks the value of the time taken for rotation (using **period_count**) against this maximum time. If the maximum time is greater than the value of **period_count**, it concludes that the machine is running; otherwise, it concludes that the machine has stopped. The values of **period_count** and **period_total** are updated by the TIMER0 overflow interrupt.

```
ISR(TIM0_OVF_vect)
{
        //Every 9.984ms
        TCNT0 = 255-78;

        period_total+=9.984;
        if(period_count<5500)
```

```
        period_count+=9.984;
        else
        period_count=5500;
}
```

The PC0 interrupt vector handles the general working of the tachometer, that is, the actual collection of the rpm and period data. It checks for a logic level low on the fourth bit of PINB, and if this low persists for 20 μs, the system assumes that one period has elapsed and the global variable **no** is incremented by one. Once the value of **no** reaches 6, that is, six periods have been completed, the system does the actual calculation of rpm and the period of rotation. If the value of the variable **period_total** is less than 360 ms, the system does nothing. Otherwise, it moves into a block where it uses the value of **period_total** to calculate the rpm and period of rotation, and turns the display on. Once out of this block, the system resets the values of **period_total** and **no** to zero again. See the code block at the top of the next page.

The tachometer displays the rpm and the period of rotation, using the function **display**. When the function is called with the value of the variable **what** being 4, the rpm is displayed, and when it is called with **what** being 5, the period of rotation is displayed. The other parts of the code set the initial values for the different variables for proper functioning of the program.

Working

To use the project, simply power it up and bring it in close proximity to a moving object, such as a fan or a motor shaft. A small magnet is attached to the fan or the motor shaft whose rotational speed is to be measured. The display will show the frequency and period of rotation.

```
if(!(PINB&(1<<4)))
{
        _delay_us(20);//Wait for disturbance to settle
        if(!(PINB&(1<<4))) //Check again if low
        {
                flag=0;//Display mode on in timeout
                no++;
                period_count=0;
                if(no==6) //6 periods are over
                {
/*******THIS BLOCK CALCULATES THE SPEED AND RPM BASED ON ACTUAL TESTING***********/
                        if((period_total/(no))<60);//do nothing if twice period less
                        than 60ms Filtering of 2000rpm
                        else
                        {
                                frequency = (12.0/period_total)*1000;
                                rpm = frequency*60;//rpm
                                period_actual = period_total/12;
                                displayon=1;//display on on running mode

                        }
                        period_total=0;no=0;
                }
        }
}
```

Project 25
Inductive Loop-based Car Detector and Counter

Ever wondered how your car approaching the traffic signal triggers a change of the light from red to green the moment you come close to the traffic lights? Or how when you enter a drive-through restaurant, a welcome message automatically plays? Well, wonder no more, because this project involves a circuit that will allow you to detect a vehicle passing through a road or a drive-through area. There are many ways to detect a car or a truck or a large vehicle. Usually, a vehicle has a large metal body, and the electrical characteristics of an inductor would be influenced and modified in the presence of such a large piece of metal. By measuring the change in the characteristics, one

could detect the presence of a vehicle. The block diagram of such a system is shown here:

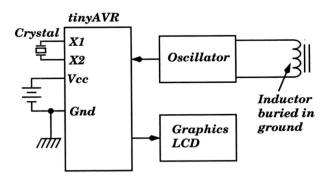

An inductor made of several turns of enameled copper wire is laid on the ground where you expect the vehicle to cross. The inductor is usually buried under the surface and covered with asphalt or concrete. The inductor is used to make a Colpitts oscillator. When a large metal object such as a car or truck passes over the inductive loop, the inductance of the coil reduces and thus the frequency of the oscillation increases. A microcontroller circuit can detect the increase in the frequency of the oscillator, and when the increase is more than a set threshold, the microcontroller can conclude that a vehicle has passed over the area.

Design Specifications

The objective of the project is to design a circuit to detect the presence of an automobile and to count the number of automobiles passing over a designated area. The system is battery powered with a graphics display to show the number of vehicles detected by the circuit.

Design Description

The system consists of two parts: a Colpitts oscillator, shown in Figure 5-9, and a microcontroller measurement and display circuit, shown in Figure 5-10. The Colpitts oscillator uses two capacitors, C1 and C2, and an inductor connected across X1-1 and X1-2 pins of the connector shown in the figure. The frequency of oscillation is $F = 1/(2\pi \sqrt{(LC)})$. C is the equivalent capacitance of C1 and C2, and is equal to: C1 × C2/(C1 + C2), that is, the series equivalent of the two capacitors. The inductor consists of several turns of copper enameled wire encased in a sealed, protective plastic covering, which is then buried in the ground. The physical size of the inductor is important and should be about six feet in length and four feet in width. Ten turns of copper wire render about 500 to 1,000μH of inductance. By choosing C1 = C2 = 1000pF, the resultant frequency of the oscillator is in the range of 200 to 300 KHz, which an AVR microcontroller can measure easily.

The output of the oscillator is fed to the microcontroller circuit shown in Figure 5-10. It

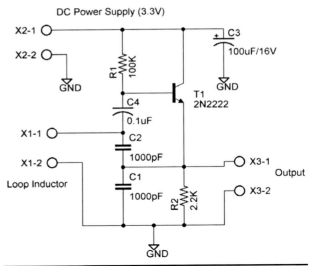

Figure 5-9 Oscillator schematic diagram

consists of a Tiny861 microcontroller with a 7.37-MHz crystal for its system clock. There is nothing magical about this frequency. It was just a handy crystal with a value less than 8 MHz we had since the maximum frequency of the Tiny861 is 10 MHz (for the version we used). The system is powered with a battery and a 3.3V LDO (LP2950-3.3V) since the graphics display (Nokia 3310) requires a 3.3V supply voltage. Other than that, the circuit has just an ISP connector to program the microcontroller.

Fabrication

The board layouts in EAGLE, along with the schematics, can be downloaded from www.avrgenius.com/tinyavr1.

The main microcontroller circuit for this project has been built using a custom PCB. The oscillator circuit has been built on a general-purpose circuit board, and the two circuits are connected together with a pair of connectors for power supply and for oscillator output. The oscillator board has another connector to connect the inductive loop. The illustrations on pages 156 and 157 show the setup of the car detector system, the car detector oscillator and controller circuit boards, and the component and solder sides of the circuit boards.

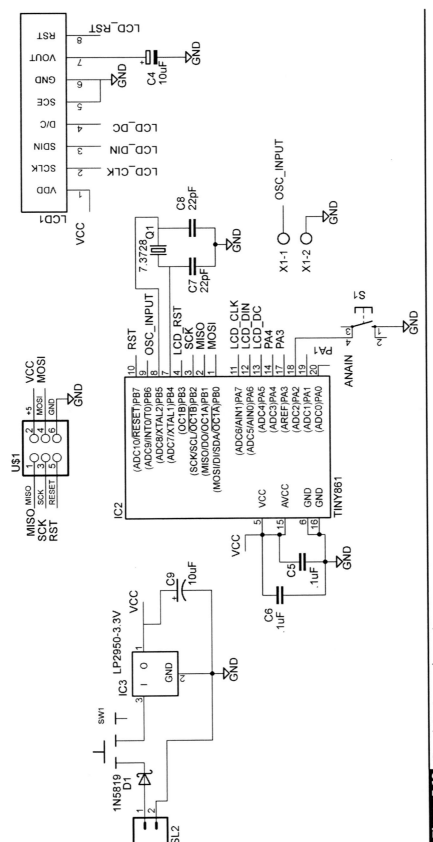

Figure 5-10 Car detector and counter

155

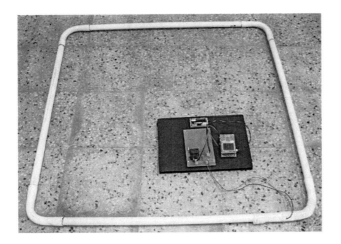

Design Code

The compiled source code, along with the MAKEFILE, can be downloaded from www.avrgenius.com/tinyavr1.

Timer 1 has been initialized in the CLEAR TIMER ON COMPARE MATCH mode. OCR1A value is set for an interrupt of 1 second in total, of which 0.1 seconds are allotted to Timer 0 for counting the external pulses coming in from the loop and 0.9 seconds system is left in the state machine.

```
TIMSK  |=  (1<<OCIE1A);                     //enable compare interrupt
TCCR1B |= ((1<<WGM12) | (1<<CS12));         //with prescaler of 256, WGM for CTC mode
OCR1A = 31250;                              //for 1sec interrupt
sei();                                      //Enable global interrupt
```

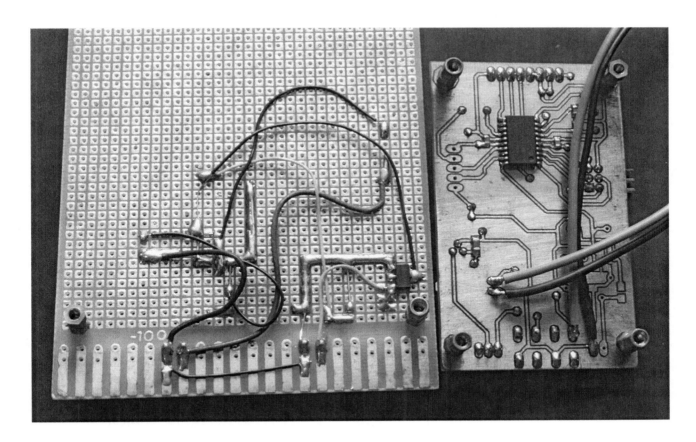

Case s0 sets the base frequency of that surface without a car when the user presses the switch 1 on the circuit board.

Case s1 keeps a check on the arrival of the car. If the changing/current frequency crosses the sum of the base frequency and the upper threshold frequency (which the user can set according to his

Case s0

```
if( !(SWITCH_PIN&(1<<2)) )        //When switch Pressed
{
base_freq = changing_freq;        //Set the Base frequency
display_base_freq(base_freq);     //display it on LCD
select_case = s1;                 //Case s1 selected
}
```

Case s1

```
while( changing_freq > (base_freq+upper_threshold_freq) )
//If changing Frequency crosses threshold
{
select_case = s2;                 //Case s2 selected
LED_PORT &= ~(1<<LED_GREEN);      //Green LED on
LED_PORT |= (1<<LED_RED);         //Red LED off
break;
}
```

Case s2

```
while( changing_freq < (base_freq+lower_threshold_freq) )
//If changing Frequency back to normal
{
        LED_PORT &= ~(1<<LED_RED);        //Red LED on
        LED_PORT |=  (1<<LED_GREEN);      //Green LED off
        car_counter = car_counter+1;      //Car counter incremented
        display_car(car_counter);         //Display number of cars
        select_case = s1;                 //Case s1 selected
        break;
}
```

or her needs in the detector.h file, initially 6 KHz), the system says that a car has arrived and turns the green LED on and the red LED off. Now the control moves to case s2.

Case s2 keeps a check on the departure of the car. If the changing/current frequency falls below the sum of the base frequency and the lower threshold frequency (which the user can again set according to his or her needs in the detector.h file, initially 2 KHz), the system says that a car has gone and turns the green LED off and the red LED on. The car counter is incremented by 1 and control goes back to case s1.

In the ISR for Timer 1 compare match, Timer 0 is initialized and set to trigger on external pulses, which are coming from the T0 pin. Timer 0 counts up whenever a rising pulse comes in.

We count the number of pulses for 0.1 second and then calculate the changing/current frequency. Then it is displayed on the graphics display screen.

```
ISR(TIMER1_COMPA_vect)
    //Setting up interrupt service
    //routine for Compare mode
{
    //Counts no. of pulses for .1 sec
    ovf_counter = 0;
    //Setting Timer 0 for external
    //pulses
    TCCR0 = ((1<<CS02) | (1<<CS01) |
(1<<CS00));
```

```
//enable Normal mode, External Clk T0
//rising trigger
//TCNT0 increases for every rising edge
        TCNT1=0;
        TCNT0=0;
        while((3125)>=(uint32_t)TCNT1)
        {
                if(TCNT0==255)
                {
                        ovf_counter += 1;
                        TCNT0=0;
                }
                counts = TCNT0;
        }

        //Total count after .1 sec
        //Calculations for finding
        //current frequency
        counts = (ovf_counter *
255) + counts;
        freq=(float)counts;
        freq=freq*10;
        counts=(uint32_t)freq;
        counts=counts/10;
        changing_freq = counts;
        LCD_partclear();
        display_changing_freq();
}
```

These macros are present in the detector.h file, which should be set by the user according to his or her needs. It has both the upper and lower threshold frequency variables that are to be used during the car's arrival and departure.

```
#define upper_threshold_freq 600
    //in khz, divide the freq by 10
//i.e It is 6 khz in above case which
//gives 6000/10 = 600
#define lower_threshold_freq 200
    //in khz, divide the freq by 10
//i.e It is 2 khz in above case which
//gives 2000/10 = 200
```

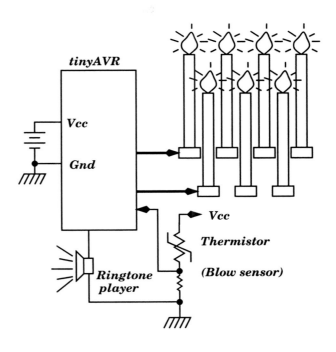

Working

To use the inductive loop-based car detector requires embedding the inductive loop under the surface. Once the loop is embedded and connected to the oscillator, the controller is turned on. The controller displays the counter value initialized to 0, and when the car approaches the loop, the display shows the presence of the car. When the car exits the loop, the counter is incremented by 1.

Project 26
Electronic Birthday Blowout Candles

We wanted to design LED-based birthday candles, except we also wanted to be able to blow them out like we would blow out a normal candle. This project describes such a candle system that is suitable for small kids since these candles are fire-free. To sense the air blow, the system uses a thermistor, as shown in the following illustration. The microcontroller controls the LEDs and can change the intensity randomly to give a perception of flickering candles. The system can be coupled with an optional ringtone player (as part of a project in the next chapter) that would play the happy birthday ringtone after all the "candles" are blown out.

Design Specifications

The objective of the project is to design LED-based birthday candles with the feature to blow out

the "candles" just like normal wax candles. The purpose of such an electronic, battery-operated solution is to create a safe fire-free solution for kids.

Design Description

The circuit diagram of the project is shown in Figure 5-11. It consists of a Tiny44 microcontroller, which has 12 I/O pins, but since one of the pins is for RESET, it actually allows only 11 pins for I/O. The circuit is powered with an external battery (four AA NiMH batteries recommended) without any on-board regulator to keep the system simple. The system has 20 LEDs in a 4 × 5 matrix. The LEDs are arranged with common anodes connected to the power supply through PNP transistors. The cathodes are connected to the microcontroller pins. These LEDs are multiplexed at a high frequency. To achieve a flickering effect, the average current through the LEDs is modulated using a software PWM scheme. The system uses a potential divider of normal resistor (R10, 150 Ohm) and a thermistor (R9, 150 Ohm nominal value). A small resistance is placed in close proximity of the thermistor so as to heat it. The voltage at the R9-R10 junction is

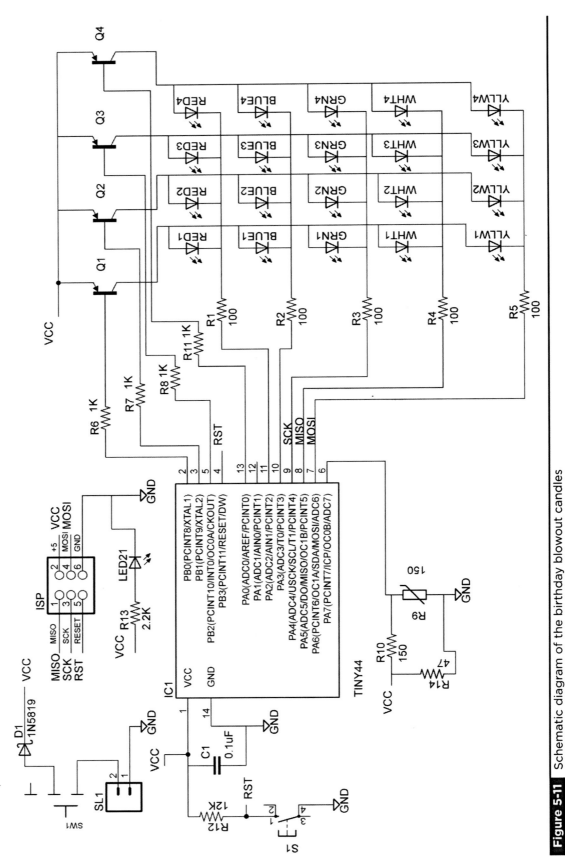

Figure 5-11 Schematic diagram of the birthday blowout candles

continuously monitored by the microcontroller analog input pin (PA7). When a person blows air on the thermistor, it cools down and this reduces the voltage at the resistor junction connected to the PA7 pin. If the voltage drop is more than a certain preprogrammed threshold, the microcontroller concludes that air is being blown on the circuit and so it starts putting off LEDs in a random sequence and also a random number of LEDs at a time. When you blow air onto normal candles, multiple candles blow out and the rest flicker, or you may fail to blow out even a single candle. We wanted to provide such a simulation with our LED candles. So by randomly blowing out a random number of candles, we can mimic the natural behavior of the candles.

One of the pins (PA1) of the microcontroller is not used in the circuit and is available for any use. This pin can be used to connect to a ringtone player, for example.

Fabrication

The board layouts in EAGLE, along with the schematics, can be downloaded from www.avrgenius.com/tinyavr1.

The board has been routed in two layers and, therefore, is not suitable for manufacturing by the Roland Modela MDX 20 machine. The circuit board has been purposefully cut in a circular shape. The following illustrations show the component and solder sides of the birthday candles project, respectively. Our website at www.avrgenius.com/tinyavr1 also has a video of the birthday blowout candles in operation.

Design Code

The compiled source code, along with the MAKEFILE, can be downloaded from www.avrgenius.com/tinyavr1.

The LED birthday candles work like a set of normal birthday candles, except with LEDs. One

can set the number of candles one wants lit (up to a maximum of 20), and in order to turn off the candles, one has to blow on them. Like with a set of traditional candles, blowing once might result in extinguishing only a few candles, and more than one blow might be required to turn off all the candles. The 20 LEDs are multiplexed using nine lines, and the multiplexing scheme is shown in the comments section of the code. The variables **e**, **e1**, **e2**, **e3**, and **e4** control the multiplexing in the

```
void setrandom(void)
{
        if(ab==0)
        {
                pwm[0]=3;pwm[1]=0;pwm[2]=0;pwm[3]=0;
                pwm[4]=0;pwm[5]=0;pwm[6]=2;pwm[7]=3;
                pwm[8]=3;pwm[9]=1;pwm[10]=1;pwm[11]=3;
                pwm[12]=3;pwm[13]=1;pwm[14]=1;pwm[15]=0;
                pwm[16]=3;pwm[17]=2;pwm[18]=4;pwm[19]=0;
        }
        if(ab==1)
        {
                pwm[0]=1;pwm[1]=1;pwm[2]=1;pwm[3]=1;
                pwm[4]=1;pwm[5]=3;pwm[6]=1;pwm[7]=2;
                pwm[8]=2;pwm[9]=3;pwm[10]=0;pwm[11]=2;
                pwm[12]=2;pwm[13]=1;pwm[14]=3;pwm[15]=1;
                pwm[16]=2;pwm[17]=1;pwm[18]=1;pwm[19]=1;
        }
        if(ab==2)
        {
                pwm[0]=2;pwm[1]=2;pwm[2]=2;pwm[3]=3;
                pwm[4]=2;pwm[5]=2;pwm[6]=0;pwm[7]=0;
                pwm[8]=1;pwm[9]=0;pwm[10]=1;pwm[11]=3;
                pwm[12]=3;pwm[13]=1;pwm[14]=0;pwm[15]=3;
                pwm[16]=1;pwm[17]=3;pwm[18]=2;pwm[19]=2;
        }
        if(ab==3)
        {
                pwm[0]=0;pwm[1]=3;pwm[2]=3;pwm[3]=2;
                pwm[4]=3;pwm[5]=1;pwm[6]=3;pwm[7]=1;
                pwm[8]=0;pwm[9]=2;pwm[10]=2;pwm[11]=0;
                pwm[12]=1;pwm[13]=1;pwm[14]=2;pwm[15]=2;
                pwm[16]=0;pwm[17]=0;pwm[18]=3;pwm[19]=3;
        }
        ab++;
        if(ab==4)
        ab=0;
}
```

software. The **setrandom()** function sets the values for the **pwm** array elements, which are accessed by the TIMER0 overflow interrupt service routine. **Pwm** array is used for giving random duty cycles to LEDs so that they flicker like real candles.

Multiplexing for LEDs is handled in the same way as shown in Chapter 3. Inside the main function, **t**, 20 LEDs have been divided into eight bundles. Each bundle is extinguished simultaneously, but bundles are extinguished in a random fashion. The array **z1** and the variable **z** are used to turn off the LEDs in a random fashion, when the user blows on the candles by generating a pseudo-random number from 1 to 8, which has not been generated in previous trials. The ADC output is used to control two variables: **present** and **past**. Depending on the difference between **present** and **past**, along with the random number generated, the appropriate bundle is switched off by equating its element locations of the **statusonoff** array to 0, which turns off the corresponding LEDs. The process of turning off the LEDs by blowing is controlled in the main function, through the block shown here:

```
if((present-past)>=3)
{
            if(z==0)
            {
                    statusonoff[0]  = 0;
                    statusonoff[5]  = 0;
                    statusonoff[9]  = 0;
                    statusonoff[12]=0;
                    z1[0] = 0;
            }
            else if(z==1)
            {
                    statusonoff[2]  = 0;
                    statusonoff[6]  = 0;
                    z1[1] = 1;

            }
            else if(z==2)
            {
                    statusonoff[1]  = 0;
                    statusonoff[7]  = 0;
                    statusonoff[14]=0;
                    z1[2] = 2;
            }
            else if(z==3)
            {
                    statusonoff[11] = 0;
                    z1[3] = 3;

            }
            else if(z==4)
            {
                    statusonoff[15] = 0;
                    statusonoff[18] = 0;
                    z1[4] = 4;

            }
            else if(z==5)
            {
                    statusonoff[10] = 0;
                    statusonoff[13] = 0;
                    z1[5] = 5;

            }
            else if(z==6)
```

(continued on next page)

```
        {
                statusonoff[3] = 0;
                statusonoff[4] = 0;
                statusonoff[8] = 0;
                z1[6] = 6;

        }
        else if(z==7)
        {
                statusonoff[17] = 0;
                statusonoff[16] = 0;
                statusonoff[19]=0;
                z1[7] = 7;

        }


}
```

Working

To use the birthday blowout candles, connect the circuit to a battery voltage (less than 5.5V) and let the heater resistance heat up the thermistor. It takes a few minutes for the thermistor to be responsive to air blow. Blow air gently on the circuit board, and watch the LEDs go out randomly.

Project 27
Fridge Alarm

This is a simple and useful project that detects and warns you if you leave your fridge door open for more than a reasonable amount of time. When you open a fridge, an internal light is turned on. If you do not close the fridge door, or if you don't close the door properly, a tiny microswitch keeps the lamp on. This project consists of a small circuit that is battery powered and has an LDR to detect light. When installed properly, the circuit inside a fridge will sound an alarm if you leave the door open for more than nine seconds. We believe nine seconds is more than sufficient for you to complete your business with a fridge! The block diagram of the project is shown next.

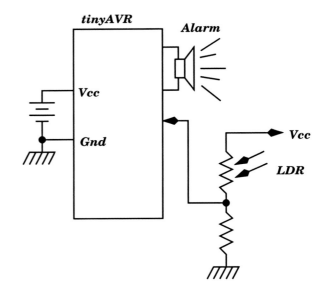

Design Specifications

The objective of the project is to design a simple, compact circuit to detect light from a fridge lamp and to measure the time for which the light is on. If the time exceeds nine seconds, then the circuit is to sound an alarm to alert the user of the fridge that perhaps the fridge door has been left open unintentionally.

Design Description

Figure 5-12 shows the schematic diagram of the project. The circuit is devoid of any protection diode and voltage regulator, so care must be taken while connecting the batteries. To keep the size of the project compact, the batteries employed are small button cell batteries, such as LR44. Four such batteries of 1.5V are connected in series to run the microcontroller at a voltage of 6V. Capacitor C1 is soldered near the supply pins of the microcontroller to decouple the noise arising in the circuit. LED1 is a 3-mm door open/close indicator green LED. The microcontroller used is the ATtiny13 microcontroller. It has one hardware PWM channel on Timer0 required for driving the transistor Q1. An 8-Ohm speaker is connected to the collector of

this transistor, as in the schematic, and a 22-Ohm current-limiting resistor is connected in series with it. The LDR is connected to the microcontroller PCINT pin with a 47-kilo-ohm resistor in series. This combination of the resistor with the LDR provides the required signal swing to interrupt the microcontroller via a pin change interrupt. Capacitor C2 is used to filter out noise spikes during the signal swing.

The source code of the project runs according to the voltage level at the pin change interrupt pin. In the presence of light, an LDR has few kilo-ohms of resistance, causing the voltage drop across it to be negligible, and this voltage serves as logic level 0. This wakes up the microcontroller from the power-down mode as the signal swing generates a pin change interrupt. The microcontroller then

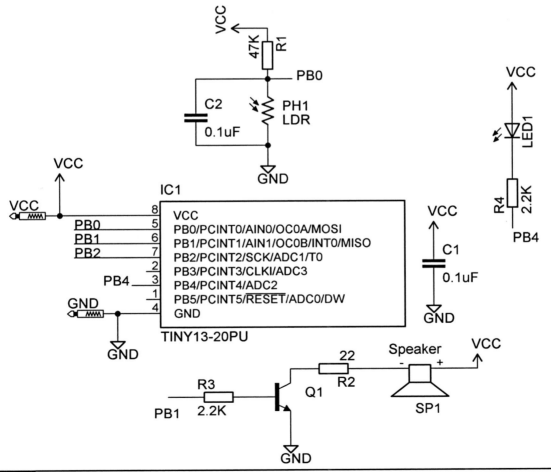

Figure 5-12 Schematic diagram of the fridge alarm

waits for a specific amount of time and if the logic level remains the same, then the alarm goes off. If the door is closed, the logic level goes high, since in the absence of light the LDR has resistance of the order of mega-ohms. This voltage across the LDR is close to logic 1 of the system; thus, this signal swing interrupts the microcontroller, again causing it to go into power-down mode.

Fabrication

The board layout in EAGLE, along with the schematic, can be downloaded from www.avrgenius.com/tinyavr1.

The board is mainly routed in the solder layer with a few jumpers in the component layer. The component and solder sides of the soldered board are shown in the following illustrations, respectively.

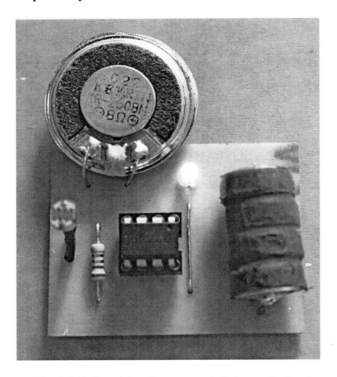

The LDR is soldered at some height such that it can be in direct contact or near the refrigerator's internal source of illumination. The illustration shows the project in an enclosure to increase its portability.

Design Code

The compiled source code, along with the MAKEFILE, can be downloaded from www.avrgenius.com/tinyavr1.

The code runs at a clock frequency of 8 MHz. The controller is programmed using STK500 in ISP programming mode. When the fridge door is closed, the controller is in power-down mode. During this mode, the controller draws only 100nA of current, thus lowering the power requirements when the door is closed and enhancing battery life. The power-down mode can be implemented by setting the SE and SM1 bits of MCUCR. The controller wakes up from power-down mode when there is an external asynchronous interrupt. This interrupt is achieved by the LDR connected on the PCINT pin. The PCINT pin generates an interrupt whenever the voltage level changes on the corresponding pin of the interrupt. The other important sections of the code are explained here:

```
while(1)
{
if(d==1)//door on after 9 secs
{
d=0;
speaker_init();//speaker initiate
TIMSK0 &=~(1<<TOIE0);
    //timer overflow off
}
if(a==1)
{
OCR0B=0x01;//frequency 1
PORTB&=~(1<<PB4);//led on
_delay_ms(200);
OCR0B=0x80;//frequency 2
PORTB|=(1<<PB4);// led off
_delay_ms(200);
}
}
```

This is the main infinite loop of the program. It consists of two if statements used to poll the two cases, which are used to control the alarm once the door is open for more than nine seconds. The first case is executed when the interrupt service routine of the timer overflow makes control variable **d** equal to 1. This if block in the main infinite loop then turns the speaker on and turns off the timer overflow interrupt. The second case is for sounding different frequencies through the speaker and blinking the LED.

```
ISR(PCINT0_vect)//pc_int routine
{
pcint_init();//enabling interrupt
sei();//setting enable bit as 1
if(!(PINB&(1)))//if PINB is 0, then a=1
{
a=0;//initial condition
TCNT0=0X00;//initialise timer
```

```
timer_init();
sei();//set interrupt enable
DDRB|=(1<<PB4);//led on
PORTB&=~(1<<PB4);
}
else if((PINB&(1))==1)
{
a=0;//initial condition
c=0;
d=0;
all_off();//all i/o off
powerdown;//go to power down
sleep_cpu();
}
}
```

This part is the interrupt service routine of the pin change interrupt, which is called each time the door is closed or opened. When the value at the pin change interrupt pin is 0, it indicates that the door is open, as the voltage drop across the LDR is close to logic level 0. This interrupt wakes up the microcontroller from the power-down mode. During this case, the timer is initialized and the alarm LED is switched on. When the door is closed, the pin logic level is 1, causing the ISR to be executed again. During this case, the control variables are reset to their initial conditions, all I/Os are tri-stated, and microcontroller is made to enter power-down mode.

Apart from this, the code includes routines that handle timer initialization, speaker initialization, and tri-stating the I/O.

Working

The LDR of the hardware is kept close to the refrigerator's internal illumination source. When the door is closed, the microcontroller is in power-down mode and draws a small amount of current. When the door is open, the hardware LED turns on, and if it is kept open for more than a specific amount of time (nine seconds), the alarm goes off.

Conclusion

In this chapter we covered the use of several types of sensors in different applications. Sensors form an important and critical component of any embedded systems project. They act as an interface between the real world and the digital world of microcontrollers. Their uses are paramount, but so are the complexities in using them. One should adhere to all the specifications and requirements mentioned in the datasheet of the sensor to be used. It's time for some music now as we move to audio-based projects in the next chapter.

CHAPTER 6

Audio Projects

IN THIS CHAPTER WE LOOK AT a few projects focusing on generating sound or interacting with the user using sound. Many projects require an audio response—for example, a short beep generated with the help of a buzzer to indicate some response to the user—but in this chapter, we look at projects where considerable effort is invested in generating the audio response.

An audio feedback in many systems is quite desirable. The feedback may indicate pass/fail or go/no-go condition. The difference between the pass and fail condition may be with a short and long beep, respectively, or with different frequencies for the two conditions. The following illustration shows how to integrate an audio response facility in any system.

The system uses a 555 Timer–based audio oscillator. The 555 Timer IC has an enable pin, which, when set to Vcc, turns the 555 on. If the enable pin is set to ground, then the oscillator is off. Another scheme that also controls the frequency of the audio signal is shown on the next page.

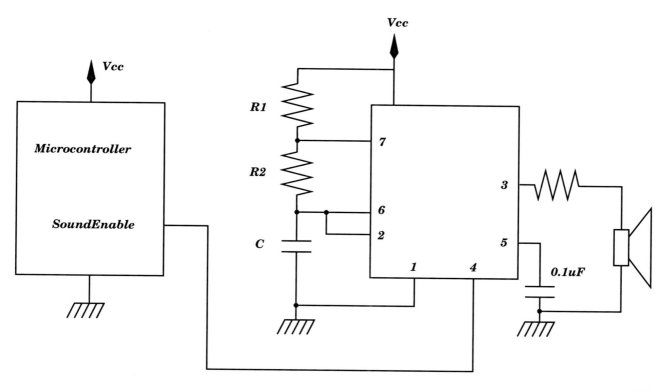

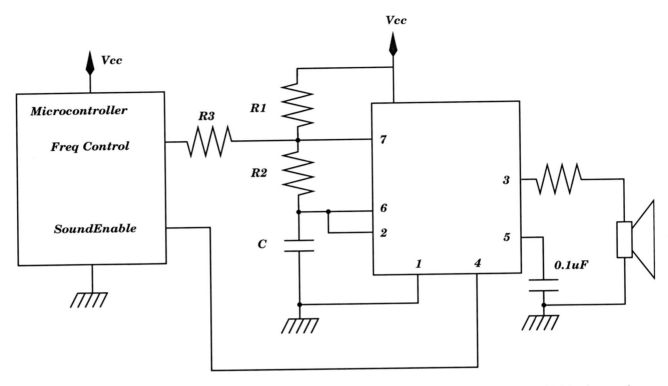

However, using an additional IC (such as a 555 Timer) in a microcontroller application is wasteful and redundant. A microcontroller is quite capable of generating audio tones. One just needs a suitable driver. The following illustration shows three schemes (a, b, and c) of generating audio signals in a microcontroller-based application. There are fixed frequency generating buzzers that just need to be turned on or off. Other options shown in the illustration include a small speaker driven directly from a microcontroller pin (option

b). Such speakers are readily available due to the boom in mobile phone usage. To drive a speaker, one needs the microcontroller to generate a square wave of the desired audio frequency (since it is easiest to generate a square wave). The square wave can be generated either using software or with the help of an internal timer. The microcontroller offers the flexibility to change the audio frequency by generating different tones for different durations.

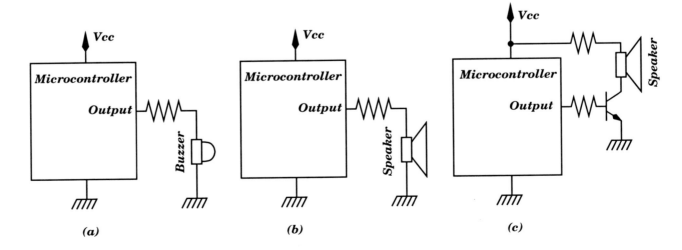

(a)   (b)   (c)

The microcontroller cannot drive too much current through the speaker, and so a current-limiting resistor must be added in series with the speaker. The resistor value should be chosen such that the total current through the speaker is less than the maximum current allowable through the microcontroller pin. On an AVR microcontroller, each pin can handle a maximum of 35 to 40mA current. The series resistance would restrict the current through the speaker, but would also mean that the sound is not that loud. If you need louder sound, the third option shown in the illustration (option c) could be used. Here, an NPN transistor is used to drive the speaker with no (or much smaller) series resistance.

The bottom illustration shows a scheme that uses an audio amplifier circuit for loud output from the speaker. Two commonly used popular audio amplifier circuits are shown. The TDA2020 would provide a really loud output suitable for use in alarm applications. The problem with these audio amplifiers, however, is the poor efficiency, since they are class B amplifiers. The illustration shows an audio amplifier based on the H-bridge that, in the parlance of audio amplifiers, would be considered class D, which has more than 90% efficiency. However, it requires more hardware features of the microcontroller to make it work.

We have used the scheme shown in the previous illustration (option c), that is, a speaker driven by an NPN transistor and the audio signal generated by the microcontroller, in a project in the previous chapter (the fridge alarm). In this chapter we look at various projects that use ready-made audio amplifier ICs as well as the H-bridge–based amplifier.

Project 28
Tone Player

This project provides a way to generate an audio signal of a required frequency. The tones are generated using the internal timer hardware of the microcontroller. The duration for which a given tone is to be played is also specified. When a

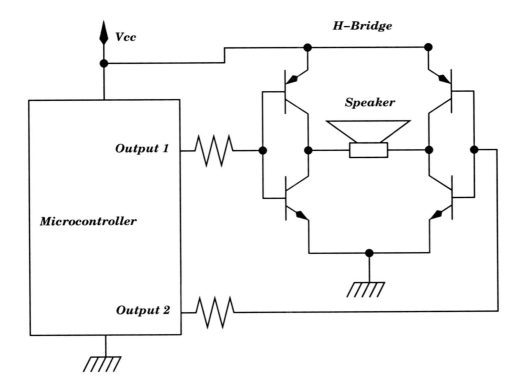

sequence of such tone-duration pairs is played by the microcontroller, it produces any required music.

Design Specifications

The objective of this project was to design a TinyAVR microcontroller-based circuit that would play music stored in the program memory of the microcontroller. The required music is specified using a sequence of tone-duration pairs. The microcontroller, when triggered, reads each pair and produces tone for the duration specified. Due to the simple nature of the arrangement, the project will not be able to produce multiple tones at the same time, as would be expected in a complex piece of music. The hardware for the project is a simple Tiny45-based circuit. The project uses a small speaker. The microcontroller drives the speaker with the help of an H-bridge circuit and produces fairly loud sound. The hardware actually consists of an LDR as well as the H-bridge audio amplifier. The project hardware is used to

implement two projects using the same common hardware, that is, the tone player as well as a modified fridge alarm. The block diagram of the hardware is shown below.

Design Description

Figure 6-1 shows the circuit diagram of the tone player and modified fridge alarm project. The four transistors on the left are the H-bridge circuit in an IC form. The H-bridge is driven by two outputs of the Tiny45 microcontroller, PB0 and PB1, which have complementary PWM outputs. The reason behind choosing complementary PWM signals to drive the H-bridge is so that when the circuit does not need to produce any sound, the two outputs are disabled (either set to "0" or "1") so that no current is drawn by the H-bridge. The output of the H-bridge is filtered by two inductors, L1 and L2, and capacitors, C3 and C4. A small speaker (8 Ohm) is connected to the output of the filters using connector SL1.

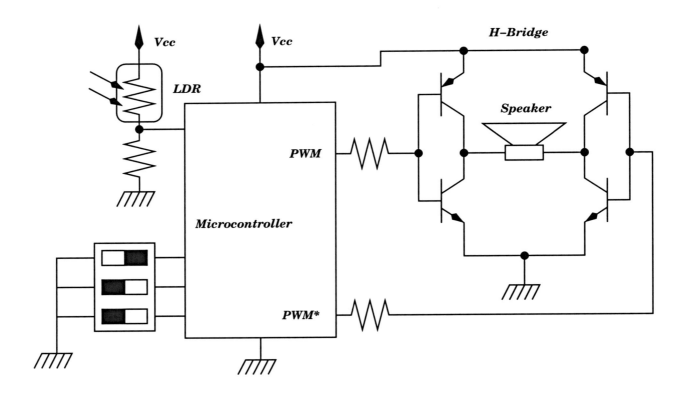

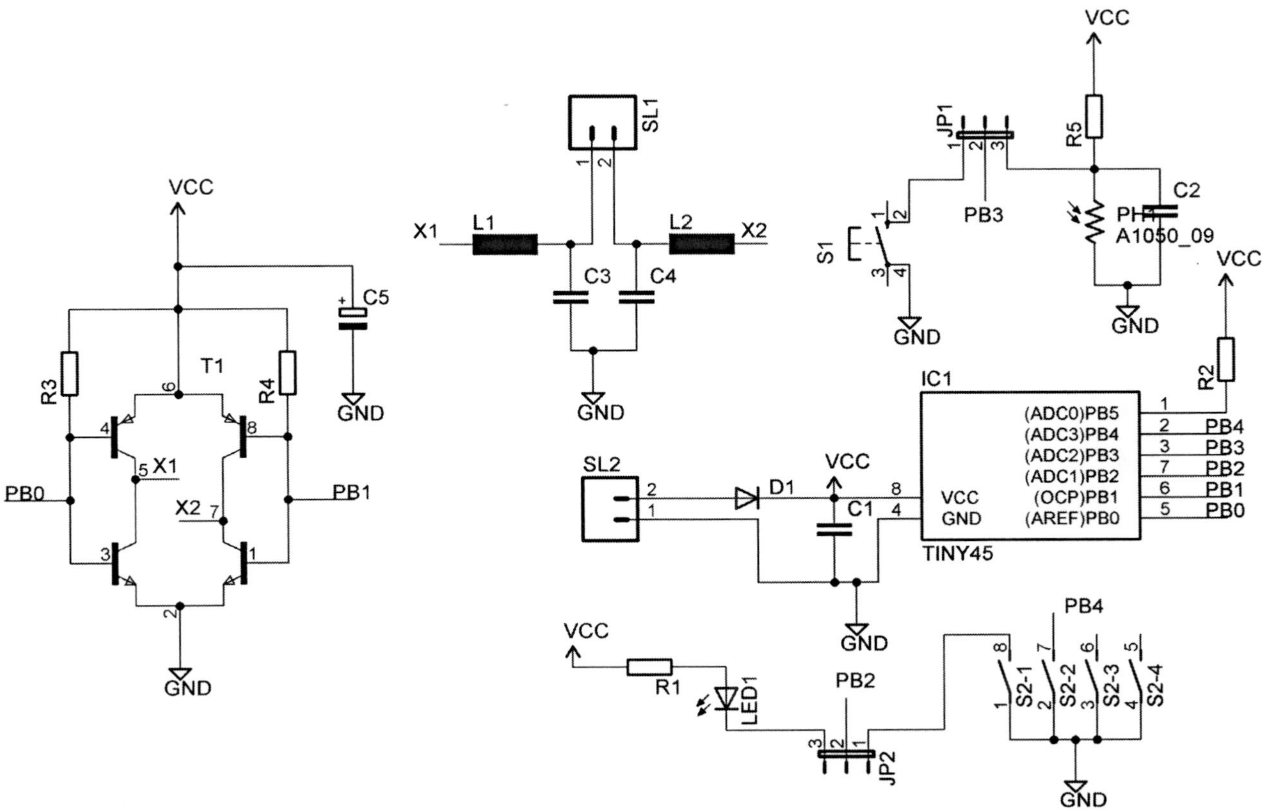

Figure 6-1 Schematic diagram of the tone player and modified fridge alarm

The pins of the microcontroller, PB2 and PB3, are shared by the two projects using jumper pins JP1 and JP2. For the tone player project, the JP1 jumper is engaged to connect the switch S1 to PB3 and the DIP switch (S2-1) to PB2. For the modified fridge alarm project, the jumpers are engaged so that LDR circuit is connected to PB3 and the LED1 is connected to PB2.

For the tone player project, the microcontroller waits for switch S1 to be pressed. Once the switch is pressed, it reads the state of the pins PB2 and PB4. PB2 and PB4 are connected to two pins of a DIP switch, which allows the user to specify one of four different data arrays stored in the program memory of the microcontroller. Each array consists of tone-duration data information that makes up a particular piece of music. The microcontroller generates the audio tone on the PB0 and PB1 pin that is connected to the H-bridge circuit, thus producing the required tone. Once the entire array

is played, it waits for the switch to be pressed again. The settings of the DIP switch can be changed to select a different piece of music before pressing the trigger switch S1.

The circuit is powered with an external DC power supply source. The voltage of the power supply should be between 3V and 6V. Using two or three alkaline batteries is a suitable source of power. Diode D1 ensures that the circuit is not damaged if the polarity of the power supply is reversed.

Fabrication

The board layout in EAGLE, along with the schematic, can be downloaded from www.avrgenius.com/tinyavr1.

The board is routed in the solder layer with a few jumpers in the component layer. The component and solder sides of the soldered board are shown in these illustrations.

Design Code

The important sections of the code are explained here. A tone consists of two parts: notes and their duration. A note represents a particular frequency, which has to be generated on the output speaker pin for a particular duration. Tone in this code has been saved using a structure with elements **duration** and **frequency** (in that order) of unsigned integer **datatype**.

```
struct _Note
{
unsigned int durationMS;
unsigned int frequency;
};
```

Songs have been restricted in the Flash memory of the microcontroller. A macro PROGMEM has been used to store the song information.

```
struct _Note song1[] PROGMEM = {
   //Birthday
{ 360, 784},{ 120, 784},{ 480, 440},
   { 480, 784},{ 480, 1048},{ 960,
   494},
{ 360, 784},{ 120, 784},{ 480, 440},
   { 480, 784},{ 480, 1176},{ 960,
   1048},{ 360, 784},
{ 120, 784},{ 480, 1568},{ 480, 1320},
   { 480, 1048},{ 480, 494},{ 480,
   440},{ 240,1396},
{ 480, 0},{ 120, 1396},{ 480, 1320},
   { 480, 1048},{ 480, 1176},{ 960,
   1048 },{0,0}
};
```

The function **play_tone_P()** takes a pointer ***p** as its argument, which traverses the **_Note** type song array, thereby giving the respective values of duration and frequency. As PROGEM has been used, we need to use a special macro **pgm_read_word** to read the flash memory. A word for the AVR is two bytes long—the same size as an int.

```
duration = pgm_read_word(p);
p++;
note = pgm_read_word(p);
p++;
top = (int)(31250/note);
```

After we get the frequency and duration of a tone, it's time to create a square wave of that frequency for the specified duration. This has been achieved by two timers. Timer0 is used in normal mode to determine the duration, while Timer1 is used in PWM mode to produce a square wave of the desired frequency by setting the TOP value of the OUTPUT COMPARE REGISTER. OCR1C acts as the top; therefore, the required frequency value needs to be fed to it. If the value of the note is zero, it means a pause. The pause is generated for the amount of time listed in the duration argument of the variable. OCR1A is the register to set the duty cycle, which has been permanently set to 50%.

```
DDRB |= ((1<<PB0)|(1<<PB1));
            TCCR0B |= ((1<<CS02)|(1<<CS00));         //1024 prescalar
            TCCR0B &= ~(1<<WGM02);                   //Normal Mode
            TCCR0A &= ~((1<<WGM00) | (1<<WGM01));
//Playing tone
            if(note)
            {
TCCR1 |= ((1<<PWM1A) | (1<<COM1A0) | (1<<CS13) | (1<<CS10);
//256 prescalar, pwm mode on
            OCR1C = top;                             //Top value
            OCR1A = (OCR1C>>1);                       //Duty Cycle
            TCNT0 = 0;
            for(;;)
            {
                    if(!(PINB&(1<<PB4)))
                    {
                            flag1 = 1;
                            return;
                    }
                    if(TCNT0 >= 78)
                    {
                            duration = duration - 10;
                            TCNT0 = 0;
                    }
                    if(duration <= 0)
                            break;
            }
            TCCR0B = 0x00;
            }
            else
            {
                    TCNT0 = 0;
            for(;;)
            {
                    if(!(PINB&(1<<PB4)))
                    {
                            flag1 = 1;
                            return;
                    }
                    if(TCNT0 >= 78)
                    {
                            duration = duration - 10;
                            TCNT0 = 0;
                    }

                    if(duration <= 0)
                            break;
            }
            TCCR0B = 0x00;
            }
```

Working

To use the circuit, you need to apply an external supply voltage. Ensure that the jumpers on JP1 and JP2 are engaged correctly. Set the DIP switches to select the required piece of music, and then press switch S1. The circuit will start playing the song. After the song is completed, it will wait for S1 to be pressed again. This circuit can be used together with the birthday blowout candles project also by connecting the pin PA1 of the birthday blowout candles circuit board to pin1 of JP1 and connecting the ground pins of the two circuit boards together. The modified code for the birthday blowout candles project (refer to our website) will trigger the tone player circuit board to play the selected song when all the candles on the candles board are blown off.

The necessary connections between the two circuit boards can be seen on www.avrgenius.com/tinyavr1.

Project 29
Fridge Alarm Redux

This project is a modification of the previous fridge alarm project. The basic concept of the project remains the same, but the alarm generation circuit has been modified. The previous project used a single transistor to drive the speaker. In the current project, we have employed an H-bridge to drive the speaker. This results in better sound performance with the same voltage requirements, but the modifications also require a different microcontroller, which is the ATtiny45.

Design Specifications

The objective remains the same as in the previous fridge alarm project, except we want to produce a louder alarm.

Design Description

Figure 6-1 shows the schematic diagram of the project. The circuit does not have any on-board voltage regulator, so care must be taken while connecting the batteries. Four AAA batteries of 1.5V are connected in series to run the microcontroller at a voltage of 6V or less. Diode D1 acts as a protection diode and protects the circuit from accidental reverse polarity. Capacitor C1 is soldered near the supply pins of the microcontroller to filter the noise. The circuit, as previously stated, was made for two projects, so some jumpers have been employed for dual use. The jumpers JP1 and JP2 are for connecting the LDR and indicator LED1 with the microcontroller. To connect the LDR and LED, pins 2 and 3 are shorted in JP1 and JP2. LED1 is a 3-mm LED used to indicate the fridge door open/close status. The microcontroller used is the ATtiny45 microcontroller. The change of microcontroller from the previous version of the project is required because Tiny45 has two complementary PWM outputs, which are required to drive the H-bridge audio amplifier. PWM channels with complementary outputs are absent in ATtiny13. These two hardware PWM channels are on Timer1. Across the output terminals of the H-bridge, a speaker is connected, as shown in the schematic, and thus current flows in both directions in the speaker. A low-pass filter arrangement is employed to filter out the high-frequency PWM signal, which acts as the carrier. This low-pass filter arrangement employs two 11-mH inductors in series and 0.1μF capacitors in parallel with the speaker. The LDR is connected to the microcontroller PCINT pin with a 47-kilo-ohm resistor in series. This combination of resistor with LDR provides the required signal swing to interrupt the microcontroller via a pin change interrupt. Capacitor C2 is used to filter out noise spikes during the signal swing.

The source code of the project runs according to the voltage level at the pin change interrupt pin. In the presence of light, the LDR has a few kilo-ohms of resistance, causing the voltage drop across it to be negligible, and this voltage serves as logic level "0." This wakes up the microcontroller from the power-down mode as the signal swing generates a pin change interrupt. The microcontroller then waits for a specific amount of time and if the logic level remains the same, then the alarm goes off. If the door is closed, the logic level goes high, since in the absence of light, the LDR has resistance of the order of mega-ohms. This voltage across the LDR is close to logic "1" of the system, thus, this signal swing interrupts the microcontroller again, causing it to go into power-down mode.

Fabrication

The board layout in EAGLE, along with the schematic, can be downloaded from www.avrgenius.com/tinyavr1.

The board is routed in the solder layer with a few jumpers in the component layer. The component and solder sides of the soldered board are shown in the "Fabrication" section of the previous project. The LDR is soldered at some height such that it can be in direct contact with the refrigerator's internal source of illumination.

Design Code

The compiled source code, along with the MAKEFILE, can be downloaded from www.avrgenius.com/tinyavr1.

The code runs at a clock frequency of 8 MHz. The controller is programmed using STK500 in ISP programming mode. The basic operation of the project and some parts of the code remain the same, except the TIMER operations. The microcontroller in use now is the ATtiny45, which has complementary PWM outputs on TIMER1. The OC1A pin and its complementary pin is used

to drive the H-bridge. The TIMER1 frequency of operation is stepped up using the internal PLL (phase-locked loop), and it works on a frequency of 64 MHz. The output PWM has a frequency of 64/256 MHz, which is about 250 Khz. Thus, a high-frequency carrier wave is generated and the sound wave is encoded using the PWM technique. The duty cycle of the high-frequency carrier wave varies according to the samples of the sound wave, and the duration for a particular sample is set according to the frequency of the sound wave. This modification is achieved using a sine wave table, which contains eight sampled values of a sine wave, as shown:

```
const char sin_wav[8] PROGMEM=
    {128, 218, 255, 218, 128, 38, 1,
    38};
```

These values are used in the main while loop of the code to encode the sine wave envelope on a high-frequency carrier, as shown here:

```
if(d_alarm)          //Loop to play tone
{
_delay_us(time);
    //time delay to play a sample
OCR1A = pgm_read_byte(sin_wav+sample);
    //Variation of Duty Cycle with
    //sinewave samples
sample++;
if( sample >= 7 )
sample = 0;
}
```

The code snippet listed runs when the control variable **d_alarm** is equal to 1. This is similar to the previous version of the fridge alarm. Inside the if statement, the **sample** variable stores the table offset and the sample value is changed by changing the compare value of TIMER1. This is achieved by changing the value stored in the **OCR1A** register as shown earlier. The time period of a particular sample is governed by the frequency of the wave to be generated. The project plays seven different frequencies for approximately

330 ms for each frequency. This is achieved using a frequency table, shown here:

```
const int sin_freq[7] PROGMEM=
   {200, 300, 400, 500, 600, 700, 800};
```

Now the change of frequency and calculation of the sample time delay is calculated in **TIMER0 ISR**, which is also used to calculate the nine-second time delay in the previous project. The code for frequency manipulation is as shown here:

```
if((d_alarm)&&(count==10))
   //Time count for tone play time
{count=0;
if(freq<7)
freq++;
else
freq=0;
freq_run=pgm_read_word((sin_freq+freq));
time= (125000/freq_run);
PORTB^=(1<<LED_PIN);
}
```

The previous if statement is a part of the **TIMER0 ISR**. The count variable is used to count the tone play time of 330 ms. After 330 ms, a timer interrupt changes the frequency that is played by changing the value of the **freq_run** variable. This variable is used to change the time variable, which governs the sample play time according to the previous formula. The time duration so developed is in microseconds and is used in the previously explained main while loop code snippet.

Working

The LDR of the hardware is kept close to the refrigerator's internal illumination source. When the door is closed, the microcontroller is in power-down mode and draws a small amount of current. When the door is open, the hardware LED turns on, and if the door is kept open for more than nine seconds, the alarm goes off.

Project 30
RTTTL Player

RTTTL (ringtone tone text transfer language) is a popular format for specifying ring tones in mobile phones. It was designed by Nokia for use in their mobile phones. As the name suggests, a ringtone is specified as a text file with codes that specify the notes and duration for each note in the ringtone. Each RTTTL file contains the name of the ringtone, the duration (d), octave (o), beats per minute (b) information, and the actual notes. This information can be decoded to play the ringtone. This project implements such a ringtone player on the Tiny861 microcontroller. The project uses a high-power audio amplifier, the TDA2020, to drive the speaker. The block diagram of the ringtone player is shown on the top of the next page.

Design Specifications

The objective of the project was to design a tinyAVR microcontroller-based RTTTL decoder and player. The purpose behind choosing the RTTTL format was the easy availability of the large number of ringtones on the Internet that could be downloaded and stored in the program memory of the microcontroller and played using this decoder.

Design Description

The hardware for this project is the same as that used in the school bell project from the previous chapter.

The schematic diagram of the project is shown here again in Figure 6-2 for easy reference. In addition, the project requires a DIP switch interface to select the required song from the microcontroller memory. The schematic diagram of the DIP switch and the pushbutton circuit is shown in Figure 6-3. The DIP switches (although four switches are shown, only three are used) allow the user to select from among eight ringtones

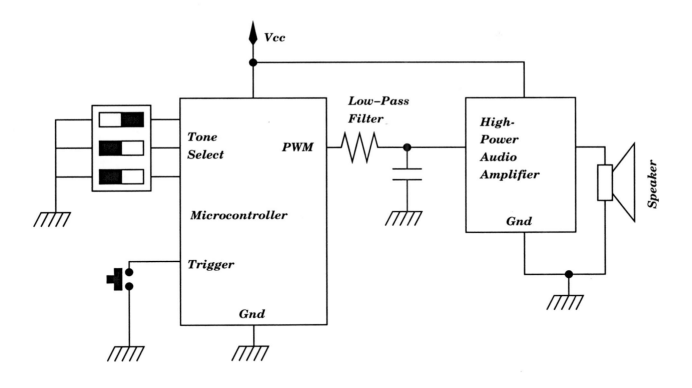

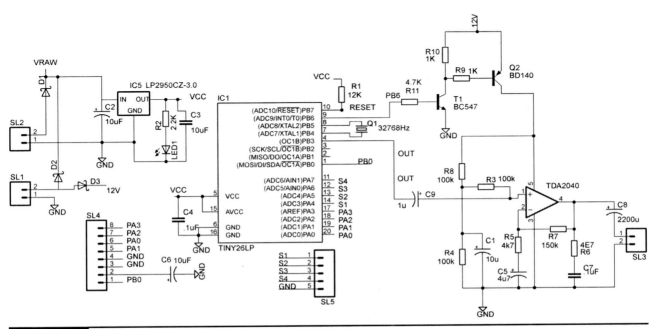

Figure 6-2 Schematic diagram of the RTTTL player

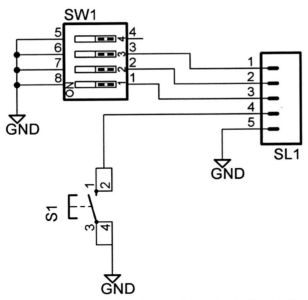

Figure 6-3 Schematic diagram of the DIP switch interface for the RTTTL player

stored in the program memory of the microcontroller. To trigger the microcontroller to play the selected ringtone, switch S1 is pressed.

Fabrication

The board layout in EAGLE, along with the schematic, can be downloaded from www.avrgenius.com/tinyavr1.

The board is routed in the solder layer with a few jumpers in the component layer. The component and solder sides of the soldered board are shown in the following illustrations.

Design Code

The compiled source code, along with the MAKEFILE, can be downloaded from www.avrgenius.com/tinyavr1.

The code runs at a clock frequency of 8 MHz. The controller is programmed using STK500 in ISP programming mode. The important sections of the code are explained here. As explained earlier, RTTTL is a text encoding containing the

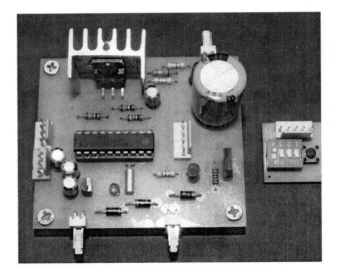

information about the characteristics of a song. Here is a sample tone in RTTTL format. At the beginning of the code, the name of song is mentioned, then the default duration (d), octave (o), and beats per minute (b) information is specified. After that, information about every note is mentioned.

```
Happy Birthday Song:d=4,o=5,b=125:8g.,
  16g,a,g,c6,2b,8g.,16g,a,g,d6,2c6,
  8g.,16g,g6,e6,c6,b,a,8f6.,16f6,e6,
  c6,d6,2c6,8g.,16g,a,g,c6,2b,8g.,16g,
  a,g,d6,2c6,8g.,16g,g6,e6,c6,b,a,
  8f6.,16f6,e6,c6,d6,2c6
```

Songs have been stored in the Flash memory of the microcontroller using the macro PROGMEM.

```
char song1[] PROGMEM = "Happy Birthday
   Song:d=4,o=5,b=125:8g.,16g,a,g,c6,
   2b,8g.,16g,a,g,d6,2c6,8g.,16g,g6,e6,
   c6,b,a,8f6.,16f6,e6,c6,d6,2c6,8g.,
   16g,a,g,c6,2b,8g.,16g,a,g,d6,2c6,
   8g.,16g,g6,e6,c6,b,a,8f6.,16f6,e6,
   c6,d6,2c6";
```

Our first task is to decode this language and to get the required information about a particular note, that is, the duration and the scale for which the note has to play. We know the frequency of a particular note, and this is stored in the array **top[]**. Here is the code for decoding the format according to the RTTTL specifications:

```
// format: d=N,o=N,b=NNN:
// find the start (skip name, etc)
while(pgm_read_byte(p) != ':')
p++;
p++;                              // skip ':'
                                  // Moving to 'd'
// get default duration
if(pgm_read_byte(p) == 'd')
{
p++;                    // skip "d"
    p++;                // skip "="
    num = 0;

while(isdigit(pgm_read_byte(p)))
    {
            num = (num * 10) + (pgm_read_byte(p++) - '0');
    }
if(num > 0)
        default_dur = num;

p++;                              // skip comma
  }
  // get default octave
  if(pgm_read_byte(p) == 'o')
  {
    p++;                          // skip "o"
    p++;                          // skip "="
    num = pgm_read_byte(p++) - '0';
    if(num >= 4 && num <=8)
        default_oct = num;
p++;                    // skip comma
  }
  // get BPM
  if(pgm_read_byte(p) == 'b')
  {
    p++;                          // skip "b="
    p++;                          // skip "b="
    num = 0;
```

(continued on next page)

```
        while(isdigit(pgm_read_byte(p)))
        {
                num = (num * 10) + (pgm_read_byte(p++) - '0');
        }
        bpm = num;
        p++;                                            // skip colon
}
// BPM usually expresses the number of quarter notes per minute
wholenote = (((60.0 * 1000.0) / (float)bpm) * 4.0);
// this is the time for whole note (in milliseconds)
// now begin note loop
while(pgm_read_byte(p))
{
        // first, get note duration, if available
        num = 0;
while(isdigit(pgm_read_byte(p)))
        {
                num = (num * 10) + (pgm_read_byte(p++) - '0');
        }
        if(num)
duration = wholenote / (float)num;          //milliseconds of the time to play the note
else
duration = wholenote / (float)default_dur;
// we will need to check if we are a dotted note after
// now get the note
        note = 0;
        switch(pgm_read_byte(p))
        {
                case 'c':
                note = 1;
                break;
                case 'd':
                note = 3;
                break;
                case 'e':
                note = 5;
                break;
                case 'f':
                note = 6;
                break;
                case 'g':
                note = 8;
                break;
                case 'a':
                note = 10;
                break;
                case 'b':
                note = 12;
                break;
```

```
                  case 'p':
                    note = 0;
      }
      p++;
  // now, get optional '#' sharp
  if(pgm_read_byte(p) == '#')
    {
       note++;
       p++;
    }


   octave = top[note];
// now, get optional '.' dotted note
if(pgm_read_byte(p) == '.')
{
       duration += duration/2;
       p++;
}
// now, get scale
if(isdigit(pgm_read_byte(p)))
{
       scale = pgm_read_byte(p) - '0';
       p++;
}
else
{
       scale = default_oct;
 }
 /* Process octave */
    switch (scale)
            {
        case 4 : /* Do noting */              // x>>y = x/2*y
             break;
        case 5 : /* %2 */
             octave = octave >> 1;
             break;
        case 6 : /* %4 */
             octave = octave >> 2;
             break;
        case 7 : /* %8 */
             octave = octave >> 4;
             break;
          case 8 : /* %16 */
             octave = octave >> 8;
             break;
            }
if(pgm_read_byte(p) == ',')
      p++;                          // skip comma for next note (or we may be at the end)
```

After we get the scale and duration of a note, we play the note for the specified duration. This is achieved by two timers, Timer0 for duration (in overflow mode) and Timer1 in PWM mode, to produce a square wave of a particular frequency by setting the TOP value of the OCR1C register.

```
DDRB |= (1<<PB3);                              //Setting the PWM channel output pin
TCCR0A &= ~(1<<WGM00);                         //Normal mode
TCCR0B |= ((1<<CS02) | (1<<CS00));             //Prescalar 1024
if(note)                                       //If a note occurs
{
TCCR1A |= ((1<<COM1B1) | (1<<PWM1B));          //Non inverting mode, Fast PWM
       TCCR1B |= ((1<<CS13) | (1<<CS10));      //Prescalar 256
       TCCR1C |= (1<<COM1B1);                  //Clear on compare match
       TCCR1D &=~((1<<WGM11) | (1<<WGM10));
       OCR1C = octave;                         //setting up Top value
       OCR1B = (OCR1C>>1);                     //50% duty cycle
       TCNT0L = 0;
       for(;;)
       {
              if(TCNT0L >= 78)                 //Duration checking
              {
                     duration = duration - 10.0;
                     TCNT0L = 0;
              }
              if(duration <= 0.00)
                     break;
       }
       TCCR0B = 0x00;
}
else                                           //If a pause occurs
{
TCNT0L = 0;
       for(;;)
       {
              if(TCNT0L >= 78)                 //Duration checking
              {
                     duration = duration - 10.0;
                     TCNT0L = 0;
              }
              if(duration <= 0.00)
                     break;
       }
       TCCR0B = 0x00;
}
```

Working

The program memory of the microcontroller is loaded with eight songs in RTTTL format. To select a particular song, the DIP switch is set accordingly and switch S1 is pressed. The microcontroller then starts playing the song. The system stops at the end of the song. To play another song (or to repeat the same song), switch S1 is pressed again.

Project 31
Musical Toy

The musical toy project is a simple musical memory game. The toy is capable of producing seven notes. In all, there are eight switches: seven switches for the notes and an extra switch to interact with the toy. To begin with, the toy produces a random note when you press the eighth switch, and the LED associated with that note also glows. Then you regenerate that note by pressing the switch associated with the note. If you guess right, the game proceeds to the next level and produces two notes, retaining the first note, followed by another note selected at random. You then generate these notes in the correct order and so on. If you fail, you can start again. If you succeed, you go to the next level. This musical toy is a good test for your musical abilities. If you can remember and regenerate a long sequence of random, uncorrelated notes, you have a musical virtuoso inside you. The block diagram of the musical toy is shown here.

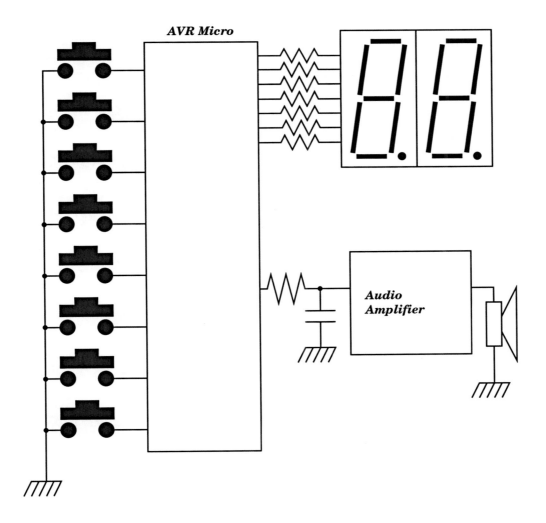

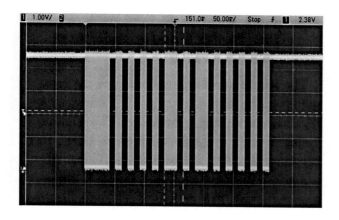

There is one major issue with IR remote devices, and that is the lack of any standard because every equipment manufacturer has their own command code formats, modulation frequency, etc. Our survey has revealed nine remote control formats: Daewoo, Samsung, Japan, Motorola, SIRCS (Sony), RC5 (Philips), Denon, NEC, and RECS80 (Thomson). Also, as devices get more and more complex, so do the remote controls. It is not uncommon to have 50 or more keys on a typical remote control. However, of

these keys, only some are used more often than others. Take a case of a TV remote control. The most common keys one uses on a regular basis are power on/off, mute, channel+, channel−, volume+, and volume−!

The remote control is such a common device that there are dedicated remote control integrated circuits. If you open any remote control device, you will find a single integrated circuit that interfaces to the keys and has an IR LED.

In this project, we decided to implement a frugal, scaled-down, batteryless TV remote that offers just the six keys mentioned earlier. The TV remote code offers user-selectable formats from NEC, SIRCS, RC5, or Samsung format for these six keys.

Design Specifications

The objective of the project was to create a batteryless TV remote control with only six keys. Figure 7-9 shows the block diagram of the

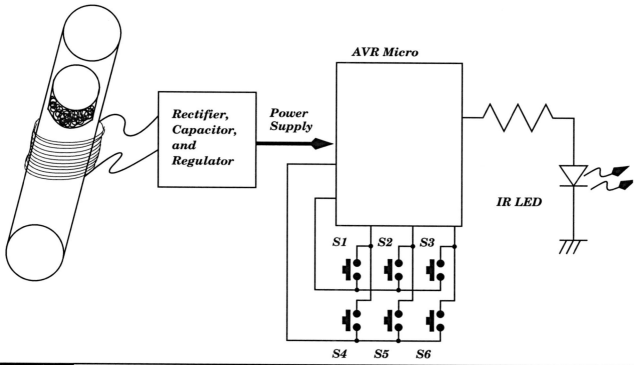

Figure 7-9   Block diagram of the batteryless TV remote

batteryless TV remote. It uses an eight-pin microcontroller. The power for operation is supplied using the Faraday generator. The six keys are interfaced using a 3 × 2 matrix with five pins. In an eight-pin microcontroller in the tinyAVR family, up to six I/O pins can be used. This project uses all six pins: five pins for the switches and one pin for the IR LED.

Design Description

The illustration shows the schematic diagram of the batteryless TV remote. The power to operate the circuit is provided by the Faraday generator. Connector SL1 in the schematic diagram connects to the Faraday generator. Diodes D1 through D4 rectify the AC voltage, and the DC voltage is filtered and stored on the capacitors C1 and C3. The voltage regulator LP2950-3.3V provides an output voltage of 3.3V to the Tiny45 microcontroller. The TV remote uses a six-

switches interface to the microcontroller on pins PB0 to PB4. Pin PB5 is used to control the IR LED in current sink mode.

The microcontroller keeps scanning the keys and upon detection of a key press, it transmits the key code corresponding to the key. The required format for each of the remote controls (REC, NEC, Samsung, or SIRCS) is programmed in the microcontroller. Although each format is different from the others, they do have some commonalities. The key code transmission begins with a start bit, followed by few address bits, and then several bits for the command code, that is, the code of the key pressed. The address refers to equipment such as a TV, audio player, DVD player, etc. In some formats, the start bit may be followed by the command code followed by the address bits. The encoding and duration of each bit also vary between the remote control formats. These details are mentioned in the code files for this project.

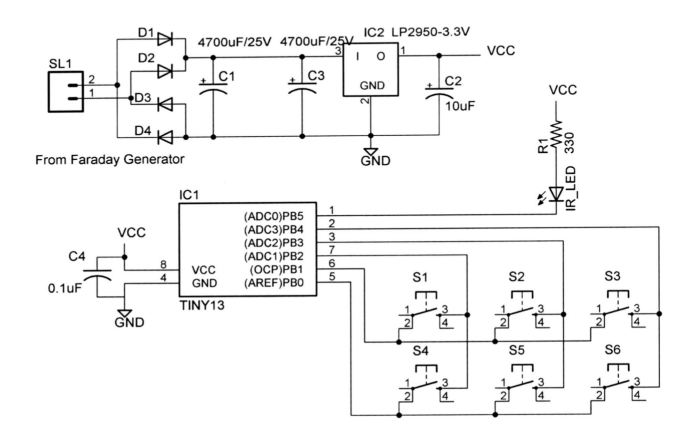

Fabrication

The circuit was built on a custom PCB and housed in a small enclosure. The Faraday generator was built to match the size of the enclosure. The following illustrations show the completed TV remote control and the insides of the enclosure.

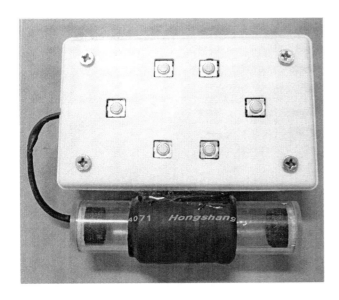

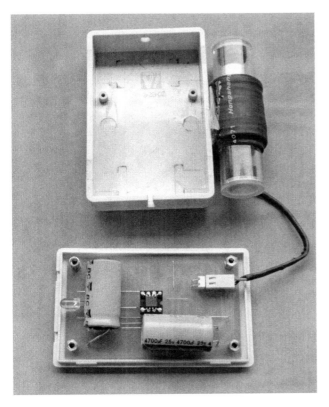

Design Code

The compiled source code, along with the MAKEFILE, can be downloaded from www.avrgenius.com/tinyavr1.

The main section of the microcontroller executes a tight loop waiting for a key to be pressed. Until a key press occurs, the microcontroller goes into sleep mode to conserve power. A key press causes a pin change interrupt, which wakes up the controller. The microcontroller executes a key scan subroutine to identify the key. In the next step, the transmit subroutine is executed to transmit the key code as per the selected protocol. Once the key code is transmitted, the microcontroller goes into sleep mode until a key is pressed again.

```
ISR(PCINT0_vect)
    //Interrupt handler for pin change
{
  MCUCR &= ~((1<<SE) | (1<<SM1));
    //Disable sleep (power down) mode
  PCMSK &= ~((1<<PCINT4) | (1<<PCINT3) |
    (1<<PCINT2));
  //Pin change interrupt is disabled on
  //all pins
  New_Key_Pressed = 1;
}
```

The key code that is to be transmitted modulates a carrier frequency. The carrier frequency depends upon the selected remote protocol. The microcontroller, which is operating at a clock frequency of 1 MHz, uses the internal eight-bit timer in clear timer on compare (CTC) mode to generate the carrier frequency. The required carrier frequency is generated by toggling the output bit. So to get 36-KHz carrier frequency, the interrupt rate has to be set to 72 KHz.

```
ISR(TIMER0_COMPA_vect)
    //Interrupt handler for compare
    //match
{
  PORTB ^= (1<<IR_LED);
    //Toggle the PIN to generate PWM
}
```

For the RC5 protocol, the timer is initialized as follows:

```
{
  TCCR0A |= (1<<WGM01);
    //Clear timer on compare mode
    //enabled
  TCCR0B |= (1<<CS00);
    //Clock frequency 8 MHz(prescalar =
    //1), CTC mode
```

```
  OCR0A = 14;
    //Approx. 72KHz Interrupt rate
  TIMSK |= (1<<OCIE0A);
    //Enable CTC interrupt
  sei();
}
```

Any key code transmission for a particular protocol involves turning off and on the IR LED at the rate of the carrier frequency modulated with the bits of the code (logic "0" or logic "1"), as shown in the illustration below.

The following code listing shows the actual bit transmission for RC5 protocol:

```
void transmit_RC5(void)
{
  while(Tx == 1)
  {
    if(Tx_bit_RC5[i] == 0)
    {
      DDRB |= (1<<IR_LED);
          //Enable carrier
      _delay_us(RC5_ON_PERIOD_ZERO);
      DDRB &= ~(1<<IR_LED);
          //Disable carrier
```

RC5 Protocol

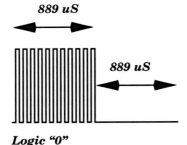

Logic "0"

Logic "1"

NEC Protocol

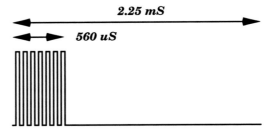

Logic "1"

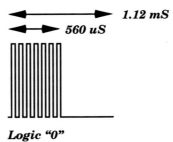

Logic "0"

```
        _delay_us(RC5_OFF_PERIOD_ZERO);
    }
    if(Tx_bit_RC5[i] == 1)
    {
      DDRB &= ~(1<<IR_LED);
              //Disable carrier
      _delay_us(RC5_OFF_PERIOD_ONE);
      DDRB |= (1<<IR_LED);
              //Enable carrier
      _delay_us(RC5_ON_PERIOD_ONE);
    }
    i++;
    if(i == 14)
    {
      i=0;
      Tx = 0;
    }
  }
  PCMSK |= ((1<<PCINT4) | (1<<PCINT3) |
    (1<<PCINT2));
}
```

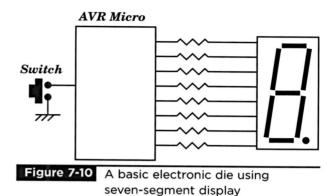

Figure 7-10 A basic electronic die using seven-segment display

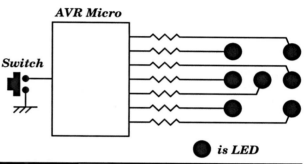

is LED

Figure 7-11 A basic electronic die using LEDs

Working

To use the batteryless TV remote is as simple as 1-2-3. Just shake the remote a few times and press the desired key!

Project 33
Batteryless Electronic Dice

Instead of traditional dice, it is nice and cool to use electronic dice. We covered RGB LED dice in a previous chapter, but let's discuss this some more. Usually, an electronic die would consist of an electronic circuit and an LED display. The LED display could be a seven-segment display that displays numbers between 1 and 6, as seen in Figure 7-10, or perhaps, to mimic the traditional dice pattern, it could consist of seven LEDs arranged as shown in Figure 7-11. Both the dice designs have a switch, which the user has to press when he or she wants to "roll the dice." The switch triggers a random number generator programmed

in the microcontroller, and the random number is then displayed on the seven-segment display or the LED display. When the user wants a new number, the switch has to be pressed again.

Both of these designs need a suitable power supply, which can be derived out of a wall wart, a suitable rectifier, smoothing capacitor, and an appropriate +5V regulator. If the user wants the dice to be portable, then the wall wart transformer should be replaced with a suitable battery, say a 9V battery. Other options for the battery exist—for example, to operate the dice from a single AA or AAA battery, a normal linear regulator will not work. To derive +5V for the dice operation, a suitable boost type DC-DC converter must be used.

But instead of using batteries or other sources of power for operation, it is possible to use a Faraday generator. Figure 7-12 shows the block diagram of such an electronic die. To generate power from a Faraday generator, it has to be shaken by moving the tube back and forth. The back and forth motion

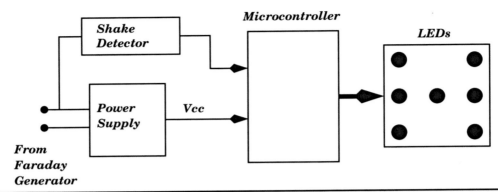

Figure 7-12 Block diagram of the batteryless electronic dice

of the tube can be detected using a "shake detector" circuit, and when one stops shaking the tube, a random number is displayed on the LED in a "traditional" dice pattern. Since the power generated lasts only as long as you shake the tube, the filter capacitor continues to provide power to the circuit for some time after the tube has stopped being shaken, and during this time, the random number is displayed on the LEDs. As the capacitor is discharged, the display turns off. To increase the time during which the LEDs remain lit after the tube has stopped being shaken, you may use a larger filter capacitor.

Design Specifications

The objective of the project is to design an electronic die using LEDs to display a random number without using any "traditional" source of power and instead to derive the operating power from a Faraday generator. Some board games require two dice. The original design is adapted to provide two sets of LED displays in the second version of the circuit.

Design Description

The illustration on the following page shows the schematic diagram of the batteryless electronic dice. Connector J1 connects to the terminals of the Faraday generator. The Faraday generator produces AC voltage and, therefore, diodes D1 through D4

(connected as a bridge rectifier) convert the AC voltage into DC voltage. The diodes 1N5819 are Schottky diodes with lower turn-on voltage compared to conventional silicon rectifier diodes. The DC voltage is filtered and stored using electrolytic capacitor C1 (4700µF/25V) and is supplied to the input of the LP2950-5V LDO regulator. The output of the LDO is 5V and is used to supply operating voltage to the microcontroller and LEDs in the circuit.

The shake detection functionality is provided by diode D5, resistor R1, and Zener diode D6. The AC input is then rectified and only positive pulses are allowed to pass through diode D5. The signal at the output of D5 is seen in Figure 7-13. The Zener diode clips the voltage pulses above 4.7V. These pulses are applied to a pin of the

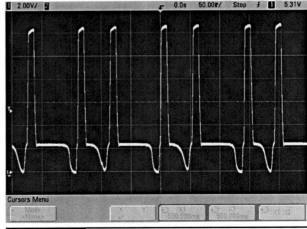

Figure 7-13 Pulse output of the shake detector

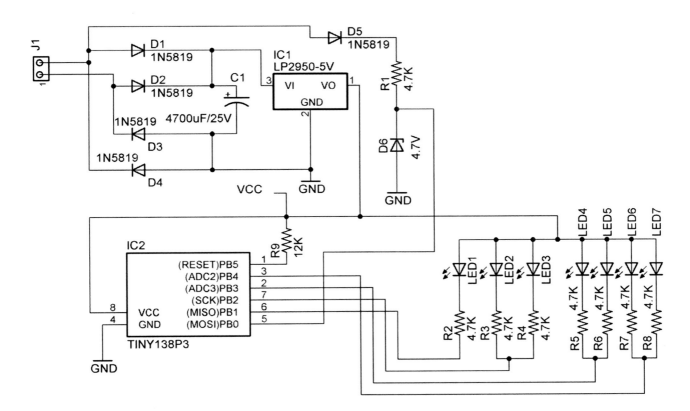

microcontroller (PB0). The program inside the microcontroller keeps monitoring the pulses, and if one stops shaking, the pulses are not produced any more. The microcontroller thus concludes that the user has stopped shaking the tube and produces a random number and displays it on the LEDs.

The LEDs are arranged in such a fashion that only four pins are required to control seven LEDs. Of course, the microcontroller does not have individual control over all the LEDs. Instead, the four pins of the microcontroller control one, two, two, and two LEDs, respectively. The LEDs labeled LED1 through LED7 are arranged in current sink mode in the fashion shown in Figure 7-14. As the user starts shaking the tube, voltage on the capacitor C1 rises, due to which the output of the voltage regulator also increases. When the output of the regulator stabilizes, it provides operating voltage to the microcontroller, which starts executing the program. The program initializes the port pins and turns off all the LEDs.

It also starts an internal timer, T0. The timer count increments on every eight clock cycles of the microcontroller system clock. The microcontroller then waits for the user to stop shaking the tube. Once the user stops shaking the tube, the microcontroller reads the timer T0 value and performs modulo 6 operation on the timer value. This results in a value between zero and five. The result from this operation (zero to five) is translated as 1 to 6 on the LED display. Once the LEDs display a random number, the available charge on the capacitor is sufficient to light the LEDs for an average time of about ten seconds. Once a number has been displayed on the LEDs, the microcontroller then waits for the user to shake the tube again. To get a new random number, the user must shake the tube a few times again.

Since the operation of the timer and the shaking of the tube are asynchronous, the resultant number from the timer is fairly random. This is how the batteryless electronic dice circuit works.

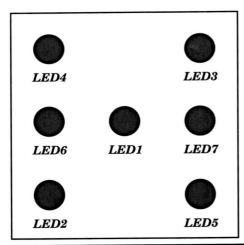

Figure 7-14 Arrangement of LEDs

Many board games require two dice, so we modified the original design to provide two sets of LED displays. The following illustration shows the schematic diagram of the dual dice. The two sets of seven LEDs are multiplexed. The program for the Tiny microcontroller is modified so that the two sets of LEDs are refreshed alternately at a high frequency so that both sets of LEDs show the random numbers. In the dual dice, the two random numbers are generated using the Timer0, as in the

single die, except the first random number is generated when the tube starts shaking and the second one after the tube stops shaking. Other than that, the dual dice operates in a manner similar to that of the single die.

For the single die as well as dual dice, the Tiny13 microcontroller is operated with an internal RC oscillator programmed to generate a 128-KHz clock signal. This is the lowest clock signal that the Tiny13 can generate internally and is chosen to minimize the current consumed by the microcontroller.

Fabrication

The first version of the single die was made on a general-purpose circuit board of about 2 cm wide and 10 cm long, as seen in the illustration on the top of the next page.

The soldered circuit is housed in another Perspex tube. After the assembly, the circuit tube and the Faraday generator tube are fixed together for ease of operation, a prototype of which is shown in the following illustration.

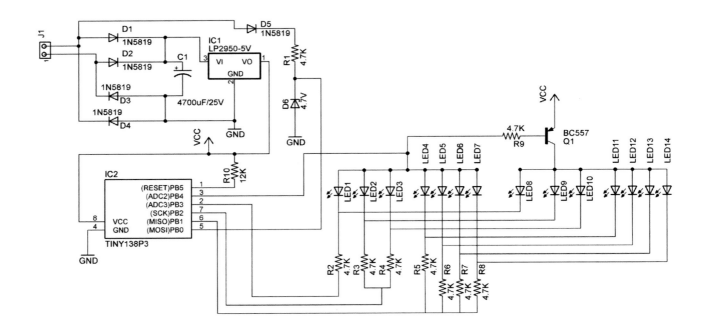

After extensive testing of the prototype, we decided to get some printed circuit boards from a vendor, and the single die and dual dice systems were soldered and packaged in the same way as the prototype, as seen in the following illustrations.

Design Code

The compiled source code, along with the MAKEFILE, can be downloaded from www.avrgenius.com/tinyavr1.

The important section of the code is the main infinite loop, where the microcontroller continuously monitors pulses on the PB0 pin. Once the pulses stop coming, it generates a random number using Timer0 and displays it on the LEDs. Similar code is available for the dual dice. In this code, we have used the **_delay_loop_2** function for delay, as opposed to **_delay_ms** and **_delay_us** that we have used so far.

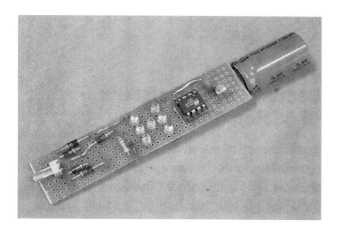

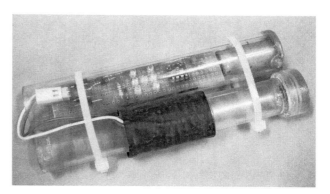

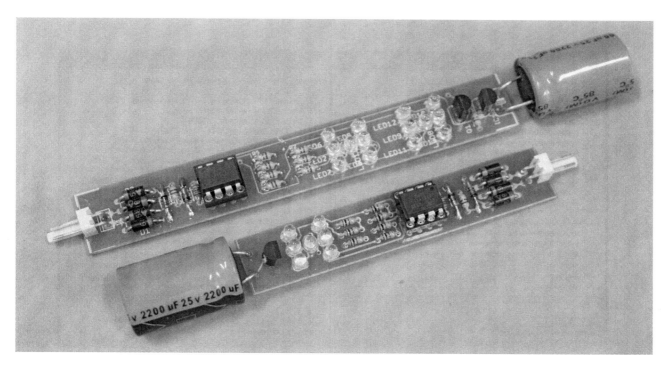

```
const char ledcode[] PROGMEM= {0xfc,
    0xee, 0xf8, 0xf2, 0xf0, 0xe2, 0xfe};
void main(void)
{
 unsigned char temp=0;
 int count=0;
 DDRB=0xfe; /*PB0 is input*/
 TCCR0B=2; /*divide by 8*/
 TCCR0A=0;
 TCNT0= 0;
 PORTB=254; /*disable all LEDs*/
 while(1)
 {
  /*wait for pulse to go high*/
  while ( (PINB & 0x01) == 0);
  _delay_loop_2(50);
  /*wait for pulse to go low*/
  while ( (PINB & 0x01) == 0x01);
  _delay_loop_2(50);
  count=5000;
  while ( (count > 0) && ((PINB &0x01)
   ==0))
  {
   count--;
  }
  if(count ==0) /* no more pulse so
   display a random number*/
```

```
  {
   PORTB=0xfe; /*all LEDs off*/
   _delay_loop_2(10000);
   temp=TCNT0;
   temp= temp%6;
   temp =pgm_read_byte(&ledcode[temp]);
   PORTB=temp;
  }
 }
}
```

The Tiny13 microcontroller is programmed using the STK500 programmer, and the fuse bits for the microcontroller are seen in the illustration.

Project 34
Batteryless Persistence-of-Vision Toy

A persistence-of-vision device uses the effect that light shown to the human eye is "remembered" after the light is turned off. The duration of time for which the eye "remembers" the light is about 8 ms. This particular characteristic of the human

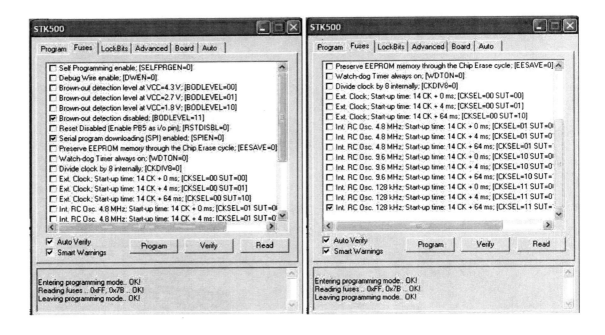

eye is employed in such toys that use LEDs to display lighting patterns spatially—that is, LEDs moving in space with changing lighting patterns. The eye remembers these patterns and is able to create a meaningful image from them. This principle was also used in the spinning LED top project in a previous chapter.

In this project, we use the persistence of vision (POV) characteristic of the eye to create a batteryless toy that displays messages in the air using a single column of LEDs. The user simply moves the toy back and forth in the air. The back-and-forth motion of the toy is used to generate the required operating voltage for the toy, as well as to change the light pattern on the LEDs. Text, as well as graphical patterns, stored in the memory of the controlling microcontroller of the toy can be displayed.

Design Specifications

The aim of the project is to design a batteryless POV toy that is programmed with text and graphics patterns in a microcontroller. The toy uses seven LEDs arranged in a column, which is waved through the air. These LEDs are controlled by the microcontroller, which generates a pattern of light on the LEDs that an external observer sees as a message or graphics due to the persistence of vision. The operating power for the toy is derived from the Faraday generator, as described earlier in the chapter. Figure 7-15 shows the block diagram of the toy.

Design Description

The illustration on the following page shows the schematic diagram of the POV toy. The diagram

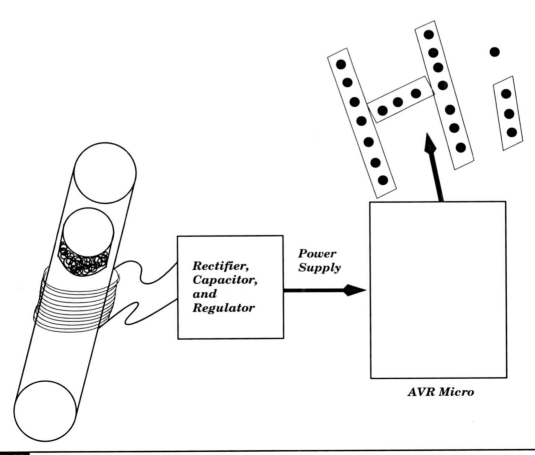

Figure 7-15 Block diagram of the POV toy

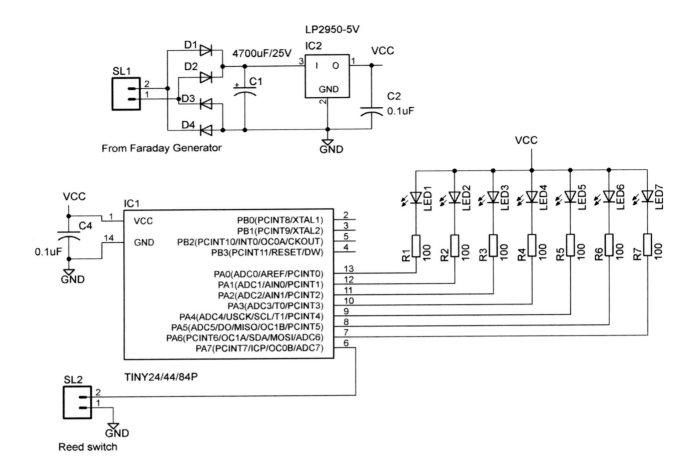

shows the bridge rectifier using diodes D1 through D4 (1N5819) and a 4700uF/25V filter capacitor. The voltage regulator used is LP2950-5V for 5V output voltage to power the microcontroller and the LEDs. The important aspect in a POV toy is the ability to generate the same LED patterns over and over in the air, and for that, some sort of synchronization signal is required. In our POV toy, this signal is generated by a reed switch connected to the tube of the Faraday generator using connector SL2. The Faraday generator is connected at right angles to the column of LEDs. As the toy is waved in the air, the magnets traverse the length of the tube. The reed switch, connected at one end of the tube, is shorted in the presence of the magnets. The magnetic field is sensed by the microcontroller and is used to synchronize the LED lighting pattern.

The LEDs are connected in current sink mode (i.e., the port pin has to be set to logic "0" for the LED to light up).

Fabrication

The POV toy was fabricated on a general-purpose circuit board, as seen in the next illustration. The circuit board was fixed on a plastic tube. Any plastic tube of suitable strength can be used. We used the tubes that semiconductor manufacturers package ICs in (incidentally, the tube we used is marked "Atmel"). At a right angle to the tube, the Faraday generator tube was fixed. After fixing the Faraday tube, we glued the reed switch and covered it with hot melt glue for extra protection. The reed switch was fixed to the right side of the tube, with the LEDs facing you.

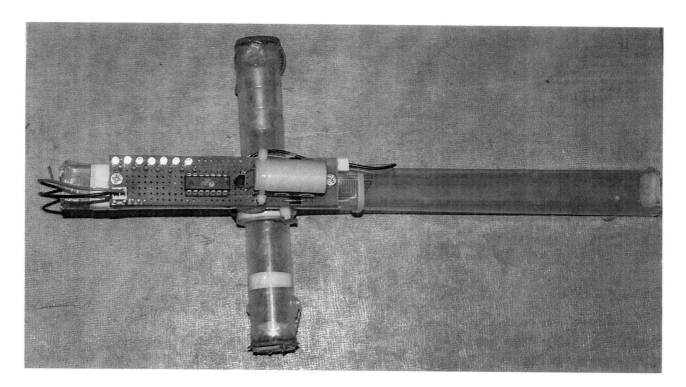

The following illustration shows a close-up view of the reed switch glued on one end of the Faraday generator tube.

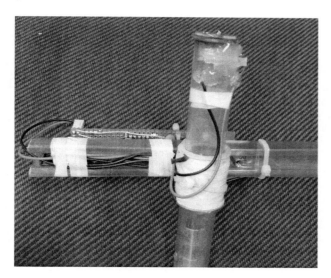

The reed switch is critical for the operation of the POV toy. A reed switch consists of two contacts of a switch coated with ferrous material. In the presence of a magnetic field, the two contacts are pulled together and the switch closes. When the magnetic field is removed the switch opens. The next illustration shows a reed switch.

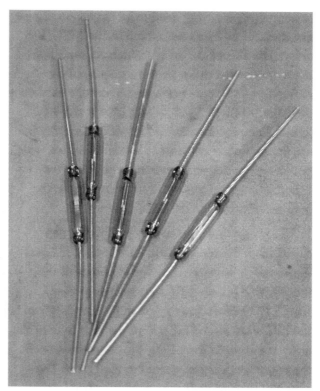

To use the POV toy, simply start shaking the toy in the air rapidly. It would start printing the programmed message as seen in the following illustration.

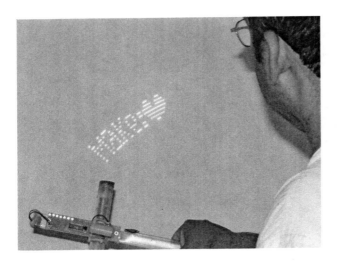

Design Code

The compiled source code, along with the MAKEFILE, can be downloaded from www.avrgenius.com/tinyavr1.

The important part of the program is the message/graphics that need to be displayed. These are encoded as bytes and stored in the **MSG[]** array. The size of the array (which depends upon the length of the message to be displayed) is defined as a constant **maxchar**. The code as shown displays "Make:♥" and so MSG has a size of 45 bytes. Since the number of entries in the array is 45, the **maxchar** constant is set to 45. If you want, you can program your own message; then the maxchar constant has to be set according to the number of entries in the **MSG[]** array.

The program waits for the synchronization signal from the reed switch, and when it receives the signal, it starts sending the elements of the array **MSG[]** to the LEDs connected to PortA. Each LED pattern lasts a few milliseconds. Then the LEDs are turned off before the next byte of the array is output to the LEDs. This continues until the entire array is transferred to the LEDs. The program then waits for the next synchronization signal from the reed switch.

```
//Make: H
const char MSG[] PROGMEM= {0x80, 0xfd,
    0xfb, 0xf7, 0xfb, 0xfd, 0x80, 0xff,
    0xdd, 0xae, 0xb6, 0xb6, 0xb9, 0xc3,
    0xbf, 0xff, 0x80, 0xf7, 0xeb, 0xdd,
    0xbe, 0xff, 0xe3, 0xcd, 0xad, 0xad,
    0xb3, 0xff, 0xff, 0x93, 0x93, 0xff,
    0xf3, 0xe1, 0xc0, 0xc0, 0xc1, 0x87,
    0x87, 0xc1, 0xc0, 0xc0, 0xe1, 0xf3,
    0xff, 0xff};
//size: 45 bytes
#define maxchar 45
void main(void)
{
  unsigned char temp;
  DDRA= 0x7f;
  PORTA=255;
  while(1)
  {
   PORTA = 255;
   while( (PINA&0x80) == 0x80);
   while( (PINA&0x80) == 0);
   _delay_loop_2(3000);
   while( (PINA&0x80) == 0);
   _delay_loop_2(1000);
   for(temp=0; temp<maxchar; temp++)
   {
    PORTA= pgm_read_byte(&MSG[temp]);
    _delay_loop_2(150);
    PORTA=0xff;
    _delay_loop_2(50);
   }
  }
}
```

Working

Using the POV toy is easy. The important task is to program the microcontroller so that it displays the desired message. The encoding of the message is shown in Figure 7-16. Port bit PA7 is used as reed switch input, while PA6 to PA0 are used to connect the LEDs in current sink mode. Thus, to turn on an LED, the port bit has to be "0." Keeping the bit D7 of the port as "1," the rest of the seven bits are encoded as shown in the figure and the message or graphics bytes are created.

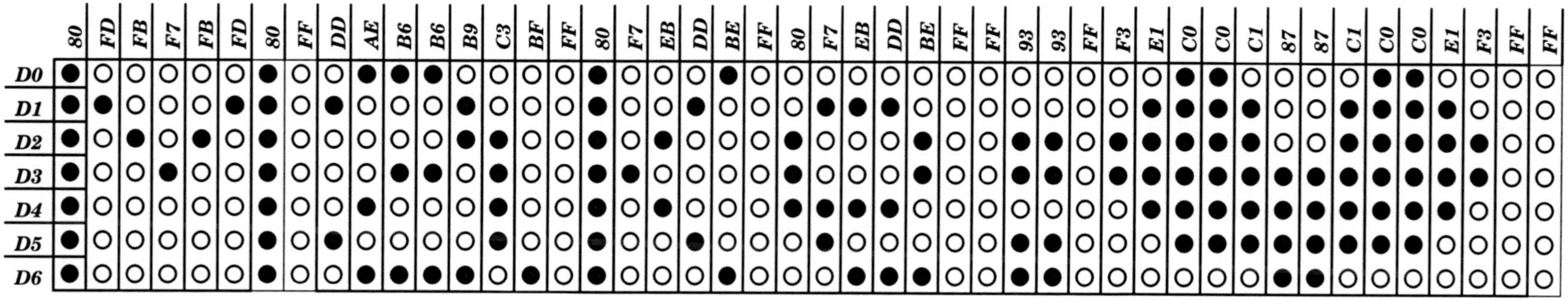

Figure 7-16 Encoding the LED lighting patterns. The message has 45 bytes.

Conclusion

In this chapter we have seen a few projects based on the Faraday generator. The basic Faraday generator can produce sufficient power for many small embedded applications, as shown in this chapter.

APPENDIX A

C Programming for AVR Microcontrollers

GONE ARE THE DAYS when one had to write the machine code for a particular function. Earlier, assembly language was primarily used for microcontroller programming. In assembly, one has to write the code in the form of mnemonics like "ADD Rd, Rs" to add the contents of register Rs to Rd. The assembler would then translate the assembly code into the machine code. But with assembly, it is difficult to write complex codes because of the low level of abstraction. Assembly is difficult to master and differs in notation from device to device.

On the other hand, C, being a high-level language, provides a higher degree of abstraction and a rich collection of libraries. C is a general-purpose programming language that can work for any microcontroller family, provided a cross-compiler exists for it. With C, one can write codes faster, create codes that are easy to understand, and achieve a good degree of software portability. C is always a popular and influential language because of its simplicity, reliability, and the ease of writing a compiler for it.

The use of embedded processors in mobile phones, digital cameras, medical plant equipment, aerospace systems, and household devices like microwave ovens and washing machines is widespread because of their low cost and ease of programming. The intelligence of these devices comes from a microcontroller and an embedded program in it. The program not only has to run fast, but also has to work in the limited memory as well. The AVR family provides many microcontrollers with different memory sizes and features. Although the programming structure for embedded devices is the same as any other software coding, it differs in the programming style. Proper optimization needs to be applied while dealing with embedded devices. One should be careful about the timing and space complexities here. It is good to avoid declaring a large number of variables. Moreover, their types should be properly selected because of space (memory) limitations in microcontrollers. For example, if a variable is always expected to have a value between 0 and 255, one should declare it as an **unsigned char** instead of **int** or **short**. Tiny devices contain limited data and program memory, so optimization needs to be done while designing systems with these devices.

Another kind of optimization needed during designing has to do with the time complexity of the codes. Timing considerations should be taken care of, especially while writing interrupt subroutines. One must have a mental picture of how one's C code is using memory and consuming time. For example, floating point calculations take more time than integer arithmetic, and should be avoided. Fixed point calculation is an alternative solution for floating point calculations.

return type, the name of the function, and the arguments it takes. After performing its task, a function may or may not return a value, depending on the return type specified. The function name can be anything, except the default keywords in C. By arguments of a function, we mean the variables that are passed to the function and that it uses to do its task.

```
int GetMax(int a, int b);
```

This statement declares a function that takes two integer-type arguments and returns a value, which is also an integer. As its name suggests, this function intends to return the maximum between "a" and "b."

■ **Definition** This contains the sequence of instructions specifying the function's behavior. It should begin with the same name and type as declared. For example:

```
int GetMax(int a, int b)
{
      if(a > b)
            return a;
      else
            return b;
}
```

■ **Calling** This is actually using the function in the program. Whenever a function is called, the control reaches to the body of the function (definition). A function can be called from any other function, and in fact, a function can even call itself (recursion). A function is called by writing its name along with the name of the variables that are supposed to be passed to it.

In C programs, there is always a function named "main" from which program execution always begins. A function can call other functions or itself as many times as it wants. It is necessary to declare the functions before they are called in a program. However, in a file, if a function is defined before it is called, then declaring that function is not necessary.

Interrupt Handling

An interrupt is a flow control mechanism that is implemented on most microcontrollers. Many events in the outside world happen asynchronously with respect to μC clock, like pressing a switch, sending a byte through a serial port, timer overflow, etc. The interrupt tells the processor about the occurrence of an event so the processor doesn't need to keep querying for it. For example, there are two ways for a processor to know if a switch is pressed or not. One way is to keep scanning the switch for its status. The other way is to tell the processor that a switch has been pressed by interrupting the execution of the main program.

When a device interrupts the CPU, the main program execution is halted and the processor jumps to a subroutine called ISR (interrupt subroutine) corresponding to that interrupt. After the required action has been taken, the interrupted program execution is resumed.

Many interrupts are allowed in AVR, some synchronous and others asynchronous. One must enable the global interrupt and the specific interrupts (maskable) that one needs to use. During interrupts and subroutine calls, the return address, which is the program counter (PC) holding the address of next instruction in the main program, is stored on the stack. The stack is effectively allocated in the general data SRAM, and consequently, the stack size is only limited by the total SRAM size and the usage of the SRAM. All user programs must initialize the SP (stack pointer) in the reset routine before subroutines or interrupts are executed.

Prototype for Interrupts

In C programming, handling interrupts is easy since there are different subroutine names for interrupts and the compiler saves the status register contents before executing an ISR. When the processor jumps to an ISR, global interrupts are disabled automatically so that no other interrupts can occur, and they are enabled at the finish of an ISR. In assembly, return from an ISR is done using the **reti** command, which enables the global interrupt enable bit and, hence, is different from the **ret** statement. The following example code shows the method to enable a timer for an AVR microcontroller for interrupt on overflow. The timer used is Timer0, which is an eight-bit timer.

Initialization of Timer0

```
//enabling the global interrupt
        sei();
//Setting timer frequency = fclk/1024
        TCCR0 = (1<<CS02) | (1<<CS01);
        TCNT0 = 0x00;
//enabling the timer overflow interrupt
        TIMSK | = (1<<TOIE0);
```

Defining the Timer Overflow Interrupt Subroutine

```
ISR(TIMER0_OVF_vect)
{
        PORTD = 0xff; //any normal command
                      //or operation
        <some more code>
}               //no need to write return
```

Every interrupt has a unique vector name to be written inside *ISR( )*, which is defined in a header file *"interrupt.h"*. Thus, one must include this header file while using interrupts and ISR name convention. The processor calls this ISR subroutine ISR(TIMER0_OVF_vect) whenever there is an overflow in Timer0, and interrupt occurs.

Arrays

Arrays are groups of similar data type elements. Arrays can be of **chars**, **ints**, **floats**, **doubles**, similar pointers, structures or unions, etc. They provide a way to declare a large number of variables without naming each of them individually. In C, an array of variables is formed by putting all of the elements contiguously in memory. This laying out of elements in contiguous memory locations is important for accessing the elements and passing them to functions. While declaring an array of elements, the size of array or number of elements should be defined with a static value. For example, an array of 100 **int** variables is declared as follows:

```
int x[100] ;
```

Here, **int** specifies the type of the variable and x is the name of the variable. The number 100 tells us how many elements of type **int** are present in this array, often called the dimension or size of the array. Any element in this array is accessed using the subscript in the square brackets ([ ]) following the array name (x here). The subscript tells us the position of the number in the array starting from 0. Thus, variables are x[0], x[1], x[2] ....x[99] and can be processed like any other normal variable. An array can be initialized with values also, like this:

```
int marks[5]    = { 90, 97, 94, 80, 91} ;
float weight[] = { 50.3, 55.4, 13.2, 20.0} ;
```

Hence, marks[2] refers to 94 and marks[3] to 80. When an array is initialized, the compiler calculates its dimension itself if not mentioned. Note that there is no method in C to determine if the subscript used in accessing an array element is out of bounds of the array. It will be out of the array memory and has unpredictable data.

More C Utilities

In this section, we discuss some additional features of C language that make it easier for the programmers to write their codes and also to improve the actual execution of the program.

The C Preprocessor

As the name suggests, this is a program that processes the source program before compilation. So, what does the preprocessor do? It processes the special commands in C programs known as preprocessor commands (sometimes called directives). These directives offer great flexibility and convenience while writing programs. This section introduces some popular and most commonly used directives in programming.

File Inclusion

Here is a directive **#include**, which allows one file to be included in another. The command for using this directive is:

```
#include "filename"
```

where filename is the name of the existing file to be included. This file should exist in the current directory of the source file or in the specified path directories. This is mostly used for including header files, although source files can also be included. Another way of using this directive is like this:

```
#include <filename>
```

with the difference that this time, the preprocessor searches for the file named filename only in the specified list of directories in the path variable.

Macro Substitution

Let's now discuss another important and widely used directive: **#define**. This is used to replace any expression, statement, or constant in the whole program. For example, see the following code:

```
#define PI 3.1415
void main ( )
{
    float circle_area, circle_circumference;
        int radius = 5;

    circle_area = PI * radius * radius;
    circle_circumference = 2 * PI * radius;
}
```

Here, the preprocessor simply replaces the token PI with 3.1415 throughout the source code. However, no substitution takes place if PI is written inside quoted strings (like "PI"). A macro can be used with an argument also or to replace a complex expression. For long expressions, one may extend it to several lines by placing a "\" at the end of each line to be continued. The following program is a valid one that uses a macro with an argument:

```
#define delay(x)  for(i=0;i<100*x;i++) \
                            asm("nop");

void main( )
{
        int i;
        ......
        ......
        delay(10);        //using macro with
                          //an argument
        ......
        ......
}
```

Now, the question is why you would use **#define** in programs. This directive makes the program easy to read, to modify, and to port. For example, a constant used many times in a program can be changed to any value without needing to change it in all the places where it is used. It is always a good programming practice to use macros.

Macros vs. Functions

As you may have noticed, macros with arguments can be used like functions. But unlike functions, macros are substituted every place where they are used without any calling and returning. This, of course, makes the program run faster, but at the cost of increased size of the output file. Hence, the concern is about space needed for a particular program. The moral of the story is to use macros instead of functions when they are small and are used many times in a program.

Macros for AVR

Writing C programs for AVR devices uses standard header files as explained earlier. Having a look at these header files, one would find a lot of macros defined for I/O pins, registers, and bit names of these registers. For example, the header file for the ATtiny45 device contains one macro for ADC. The ADCSR register has eight bits with positions from 0 (LSB) to 7 (MSB). These are defined in the header file and can be used readily in a C program.

Enumerated Data Types

An enumeration is a data type consisting of a set of values that are called integral constants. In addition to providing a way of defining and grouping sets of integral constants, enumerations are useful for variables that have a small number of possible values. The general syntax for defining an enumerated data type in C is:

```
enum Days{
    Sunday = 0,
    Monday,
    Tuesday,
    Wednesday,
    Thursday,
    Friday,
    Saturday
};
```

Here, Sunday has value "0" and the consecutive days have values of increasing order, that is, Monday = 1, Tuesday = 2, and so on. One can assign other values to these names explicitly, like:

```
enum cards{
    CLUBS      = 1,
    DIAMONDS   = 2,
    HEARTS     = 4,
    SPADES     = 8
    };
```

Volatile Qualifier

Types of data can also be qualified by using a type qualifier. One such type qualifier is *volatile*. When a variable is declared, the compiler puts certain optimizations on it according to the situation. The purpose of volatile is to force an implementation to suppress optimization that could otherwise occur. For example, the compiler sometimes loads a variable from data memory to its registers and performs some operations on it. Now the changed value is not written back to memory instantly. So when any other subroutine uses that variable, it gets the old value, resulting in a wrong output. A variable is qualified with volatile as shown:

```
volatile int temperature;
```

It is best to use the volatile qualifier when sharing of variables occurs, like with one global variable among different functions and interrupt subroutines.

Const Qualifier

Another qualifier available in C is *const*, which makes objects constant by placing them in read-only memory. A constant object should be initialized at the time of declaration. The syntax is:

```
const int my_marks;
```

This appendix provided a brief discussion on C concepts required for AVR microcontrollers. For in-depth coverage of the topic, readers may refer to any C programming book.

Designing and Fabricating PCBs

In Chapter 1, we explained the advantages of making a custom PCB (printed circuit board) for our projects over using a general-purpose board. We also discussed the several different types of software available for designing PCBs; out of those, we have used the free version of EAGLE (Easily Applicable Graphical Layout Editor) from CadSoft. There are three stages of PCB design, namely schematic design, layout design, and routing. Layout design and routing are often customized for the fabrication process that has to be followed to make the board after designing it on a PC. We used the Roland Modela MDX-20 PCB milling machine for manufacturing our boards, and so we have done the design rule check (DRC) settings in EAGLE according to this machine. The settings for different projects are largely the same with minor variations.

EAGLE Light Edition

CadSoft offers three different versions of EAGLE: Professional, Standard, and Light. These differ from each other on the basis of the maximum number of sheets in the schematic, signal layers in routing, and maximum board area for layout. We have used the Light edition to make our boards. This version has the same features as the Professional edition, but their usage is bounded within certain limits. It offers Schematic Editor with only one sheet allowed, Layout Editor with a maximum allowed board size of 4 by 3.2 inches,

and Autorouter capable of routing tracks in only two signal layers (top and bottom). These features are sufficient for the PCBs of all the projects discussed in this book. A PCB designed in the Professional edition can still be viewed in the Light edition but can't be edited. CadSoft allows the free use of EAGLE's Light edition for noncommercial projects. You can download the latest version from http://cadsoft.de. CadSoft updates their software frequently. In fact, this book was started with version 5.6 of EAGLE, and now at the time of this writing, version 5.10 is available.

EAGLE Windows

There are three main windows/GUIs that you would require for schematic entry and designing a PCB using EAGLE: Control Panel, Schematic Editor, and Layout Editor. Each is introduced below.

Control Panel

The Control Panel is the first window that appears after starting EAGLE. Its screenshot is shown in Figure B-1.

The Control Panel is the central system for the software and gives commands to other windows for their operations. To start a new project in EAGLE, go to File | New | Project. This creates a new folder in the "eagle" directory. The "eagle" directory is the default directory for storing

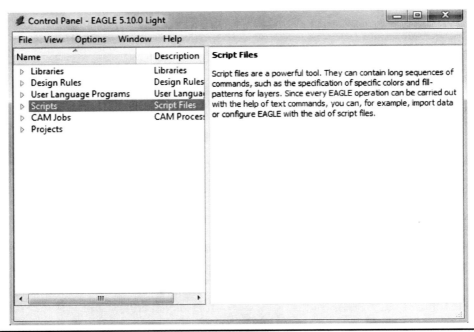

Figure B-1 Control Panel

projects, and its location is specified while installing EAGLE. After that, you can create schematic and board files under it by going to File | New | Schematic/Board. When saving the schematic and board files for the first time, you need to specify the destination directory, which, by default, is the project directory created earlier. However, you can store these files at a different location. Generally, a new board is not created as such. First of all, a new schematic is made, and after designing the circuit, the board is made from the schematic itself.

Schematic Editor

This is the window for designing schematic (circuit) diagrams, and it opens up when you create a new schematic or load an existing one. Its screenshot is shown in Figure B-2.

Layout Editor

This is the window for placing the components on the board and routing the tracks. Eagle keeps the schematic and board in sync if both windows are

open simultaneously, that is, a change in the schematic is reflected in the board also. This is called forward and back annotation. However, if the layout window is not open while editing the schematic, this sync is lost and EAGLE can't keep track of further changes. It becomes virtually impossible to design the board then, and the only option is to make a new board from the schematic. A screenshot of the Layout Editor is shown in Figure B-3.

EAGLE Tutorial

The instructions related to EAGLE commands, settings, adding components to the schematic, laying out the board, routing (manual/auto), and other things required for designing the board are discussed in detail in the tutorial that is provided from CadSoft in the EAGLE package itself. The document can be found at Installation Directory\EAGLE-5.10.0\doc\tutorial-en.pdf. It covers all the prerequisites for designing the boards of the projects discussed in this book. Also, if some of the terminology pertaining to EAGLE

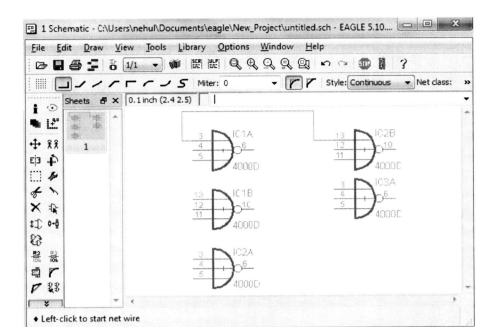

Figure B-2 Schematic Editor

that has been used in this appendix up until now is not understood, it would become clear if you read this document. A more comprehensive manual can be found at Installation Directory\EAGLE-5.10.0\doc\manual-en.pdf. You are advised to read the tutorial before proceeding to the next section.

Adding New Libraries

The packages for some of the components used in this book have been made by us and packed in a library. To use these components, you need to add the library to the EAGLE Control Panel. The

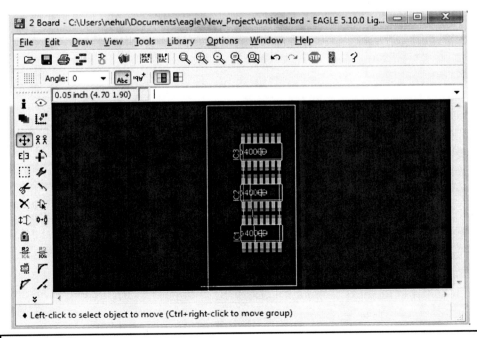

Figure B-3 Layout Editor

library file can be downloaded from www.avrgenius.com/tinyavr1.

You can save this file in any directory, but it is best to put this file at Installation Directory\ EAGLE-5.10.0\lbr, which is the default storage directory for EAGLE library files. Once you have saved this file, go to the Schematic Editor and click the Library menu in the top menu bar. Then select Use and select the previously saved library file. After this, you can add components from this library to your schematic and board, as explained in the tutorial document.

Placing the Components and Routing

Before you start laying out the board, change the DRC settings according to your fabrication process. Once this is done, place the components, keeping in mind the limitations and boundaries of the fabrication process. If your fabrication process has a poor resolution, your components need to be placed far away from each other.

Routing can be done either manually or using EAGLE's auto-routing feature. If you are designing a single-sided board, auto-routing would not be able to route all of the tracks in most cases, so they have to be done manually, either in the same layer or in the other layer. When the tracks are routed in the other layer, they are called jumpers and are not put on the board by the fabrication equipment. Instead, the nodes that are connected by this track are done so either by using tinned copper wire or insulated copper wire. These tracks can also be left unrouted because they have to be made through external wires in any case, but if jumpers are small in length, straight, and not passing through any component, the final soldered board looks clean. We describe the fabrication process we used in detail in the next section.

Roland Modela MDX-20 PCB Milling Machine

Now, we move to the fabrication of PCBs using the Roland Modela MDX-20 PCB milling machine. We will describe how to fabricate a single-layered printed circuit board (routed in the bottom layer of EAGLE) with through-hole components placed on the opposite side of the routed layer and SMD (surface mount device) components placed on the same layer, which is tracked or routed. The software used in this process has been tested on Windows. One piece of software, CAM.py, is not able to send commands to the serial port on Windows because it has been written for Linux. However, its output file can be sent to the serial port through different software, as described later.

Step 1: Making the Schematic and Laying Out the Board in EAGLE

A basic knowledge of EAGLE Schematic Editor and Layout Editor is a prerequisite for this. So you are advised to go through the tutorial mentioned in earlier sections, if you have not done that yet. The first step is to design the schematic according to your requirements. Once your schematic is complete and error free (use the menu command Tools | Erc to verify this), you are ready to design the PCB layout. In the Schematic Editor, go to File and click Switch To Board. The Layout Editor will open. First of all, make sure that the outer boundary of your board starts at the coordinates (0,0). This is required to avoid confusion when you specify the offsets. Place the components any way you want.

To make sure the board can be fabricated with the Roland Modela milling machine, we will use a set of design rules that specify the layers used (the bottom layer only), the minimum distances between pads and traces, diameters of holes, width of copper traces, etc. Open the Design Rules dialog with the menu command Tools | Drc and load the modela.dru file (provided at www.avrgenius.com/tinyavr1) using the Load button.

With the design rules loaded, lay out the PCB tracks, either manually or with the help of EAGLE's auto-router. As mentioned before, it may not be possible to route all the tracks in a single side, so you can use jumpers or leave them unrouted. When the layout is complete, verify it using the menu command Tools | Drc | Check. If it shows errors, remove them by changing the position of components, tracks, etc.

Step 2: Creating the Toolpath for Drilling

From the original PCB layout, we can prepare data for cutting and drilling with the Modela machine. In this step, we will create the toolpath for drilling the holes of the PCB.

In order to create the outline of the signal traces and holes, there is an EAGLE User Language Program (ULP) written by Marc Boon (fablab mill-n-drill.ulp) that will create the toolpath for milling the outlines of the tracks and drilling the holes. However, we will not be using this data for milling the outlines of tracks. To mill the tracks, we can use the already routed tracks in the bottom layer along with pads and vias in their respective layers. The modified version of the ULP program, customized for our requirements, can be downloaded from www.avrgenius.com/tinyavr1.

Before running the program, you have to create two new layers in the Layout Editor:

- Layer no 111 Name: Roland_Milling
- Layer no 112 Name: Roland_Drilling

To create the layers, click the layer icon in the command toolbar, and click the New button. After you have created the layers, run the ULP program by selecting File | Run and then open fablab-mill-n-drill.ulp. You should see a dialog as shown in Figure B-4. Specify the tool diameter (0.79375 mm is recommended, as it corresponds

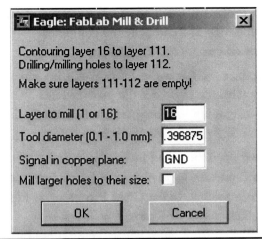

Figure B-4 Fablab mill-n-drill.ulp screenshot

to a 1/32-inch diameter drill bit) of the milling tool mounted in the milling machine for drilling the holes. Next, specify the signal that should *not* be isolated from the copper plane. By default, this is the GND signal. If you want all signals to be isolated, make sure this field is empty.

Finally, there is a check box that should be selected if you want holes that are larger than the tool diameter to be milled to their specified size. If you don't select this, all holes will be drilled and will have the diameter of the tool. Clicking the OK button will create the toolpath for milling in layer 111 and the tool path for drilling in layer 112. As explained before, we would only be using layer 112.

Step 3: Creating Drilling and Cutting Files for Driving the Roland Modela Milling Machine

To create the computer numerical control (CNC) files that will drive the Roland Modela milling machine, we will use EAGLE's CAM Processor. The CAM Processor can create output files for a variety of plotter and printing devices. The Roland Modela is not one of them, however, but we can define it ourselves by specifying the required commands for this machine in a file called eagle.def, located in the bin folder of the EAGLE installation. By adding a few lines to this file, the

CAM Processor now knows this machine as an output device and can generate computer numerical control (CNC) files for it.

Replace the file eagle.def in EAGLE's bin folder with the eagle.def file provided at www.avrgenius.com/tinyavr1, and restart EAGLE after saving your board and schematic.

Now open the saved layout again and start the CAM Processor by clicking the fourth button on the main toolbar. In the CAM Processor, load the job fablab mill-n-drill.cam, which can be downloaded from our www.avrgenius.com/tinyavr1 using the menu command File | Open | Job. You will get a dialog as shown in Figure B-5.

This dialog contains the definition of this CAM job. There are two sections: Bottom Copper Contour Milling and Hole Drilling Bottom. Each section specifies a device (from the modified eagle.def), an output file, some options, and the layers used.

All settings are predefined in the CAM job and should not be modified. The only exceptions are

the X and Y offsets in the lower-left corner. Since we are milling the bottom side, the layout is mirrored (hence, the Mirror check box is selected). However, mirroring the layout means that all coordinates will be mirrored around the Y-axis, translating positive X-coordinates to negative ones. We will have to offset our layout so it will be in the range of the milling machine's coordinate system. Selecting the pos. Coord check box is supposed to do this, but it doesn't do it properly (it adds too much offset). So we leave this unchecked and specify the offset manually.

Specifying the Offsets (Important)

There are lots of issues involved with specifying the offsets. First, the layout is mirrored, which gives negative X-coordinates, and also we don't want our board to be milled at coordinates (0,0) because it corresponds to the corner of the raw board and may be damaged or deformed at that point. Hence, it is always a good idea to start your board at the coordinates (1,1) specified in inches.

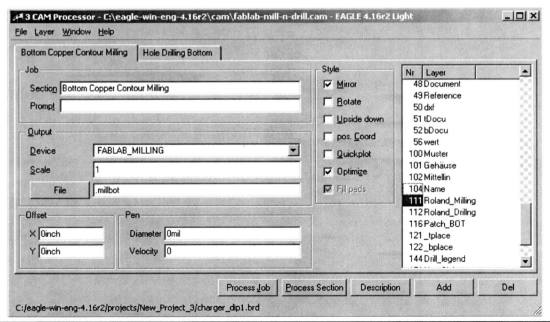

Figure B-5 fablab mill-n-drill.cam screenshot

Now go to Layout Editor and take the following readings:

- Rightmost X-coordinate (gives the width of the board)
- Lowermost Y-coordinate

As we mentioned earlier, if you have placed your board at (0,0), the lowermost Y-coordinate will be 0. We have taken the rightmost X-coordinate, because when the board is mirrored its right side becomes the left side and vice versa. Now offsets are calculated as:

- X offset = Rightmost X-coordinate + 1 (in inches)
- Y offset = 1 – Lowermost Y-coordinate (in inches)

Specify these offsets in both sections. Make sure that the layer selected in the Bottom Copper Contour Milling section is 20 – Dimension and the layer selected in the Hole Drilling Bottom section is 112 – Roland_Drilling. Click the Process Job button. It will create two files with extensions

.millbot and .drillbot in the same folder that contains your .brd files. The file with the extension .drillbot is used for drilling the holes, and the file with the extension .millbot is used for cutting the board.

Step 4: Creating Milling Files for Driving the Roland Modela Milling Machine

As mentioned in step 2, we use EAGLE's routed tracks, pads, and vias to generate the milling data. Unfortunately, there is no single job available, as in step 3, that can generate the milling data that can be read directly by the machine. So we follow a two-step process for this. Go to the CAM Processor and load the job gerb274x.cam, which is present in EAGLE by default. You will get a dialog as shown in Figure B-6.

This dialog contains the definition of this job. It contains five sections, but for single-layer (bottom) boards that have to be made without any identification layer, only a single section—Solder Side—is useful. Make sure that the four layers

Figure B-6 gerb274x.cam screenshot

shown in the previous figures are selected. The rest of the settings should not be changed from their default configuration. Specifying offsets here is not important because we would do it in the second step. By clicking Process Job, you would get multiple output files, but only the file with the extension .sol is required.

The .sol file is not readable by the Modela machine and has to go through one more stage of transformation. For this we use software called CAM.py, which is written in Python and, hence, is platform independent. However, it cannot capture the serial port of a Windows PC. The file CAM.py can be downloaded from www.avrgenius.com/tinyavr1.

Cam.py (translation: computer-aided machining file written in the Python programming language) tells Modela what to cut and how to cut, and this clever software tool was developed by Prof. Gershenfeld. This software tool was developed in Python and runs on all platforms. To run this software on Windows, the computer must have preinstalled Python, or you can install Python from www.avrgenius.com/tinyavr1. The software cam.py also requires a library called Python Imaging Library (PIL). To date, this library is available for Python versions below Python 2.7, so the installed Python version must be below Python 2.7. The library PIL can be downloaded from www.avrgenius.com/tinyavr1.

The software cam.py can now be operated on a PC running Windows by executing the cam.py file from the Python command line. The execution also can be made from the system's command prompt. Save cam.py in the same directory where Python is installed. Open a command prompt window, and navigate to the Python directory. Then type **python cam.py** on the command prompt window, and the

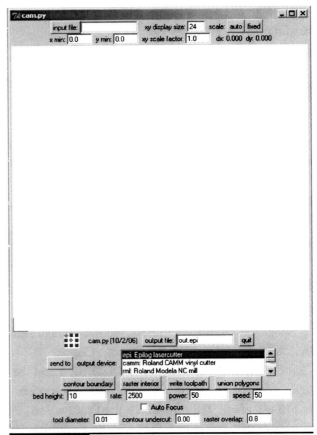

Figure B-7 Cam.py screenshot

software window will open. Figure B-7 shows a screenshot of the cam.py window.

At the top of the cam.py window, you can see some user controls, as shown in Figure B-8. These are described next.

Click the Input File button, navigate to your .sol file, and open it. You will see your layout in tiny form in the bottom-left corner of the window. You can make it larger by either changing the number in the xy display size box or you can click the next box, Auto, and the file will just about fill your window. This is shown in Figure B-9.

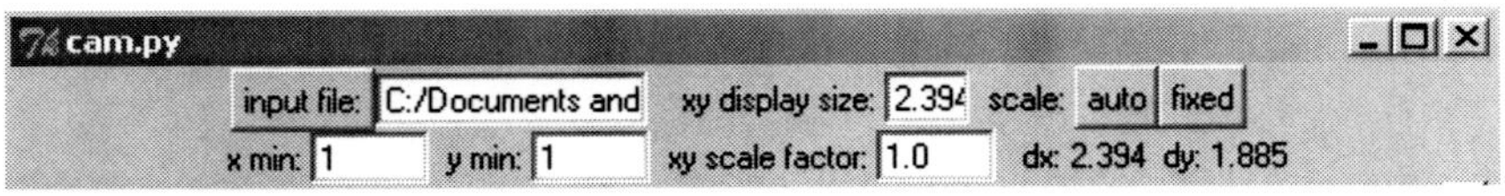

Figure B-8 Cam.py user controls

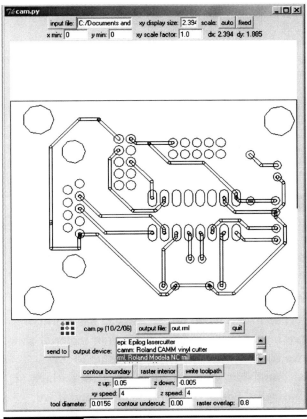

Figure B-9 Cam.py loading a file

from (1,1) inches. So, change x min and y min to 1 and 1. The next parameter is the xy scale factor, and this should be set to 1 if the board has to be the same size as that in EAGLE. Next to the scale factor are dx: and dy: These tell you the size of your imported object. At the bottom of the CAM window, there are other buttons, as shown in Figure B-10.

To determine what machine you are using and the file extension name, look at the Output Device menu below the Output File button. By selecting one of the different machines, you automatically pull up a set of tool parameters that relate to that machine, and if you look in the Output File window, you'll see that a file with the correct extension attached to it has been created. Select Roland Modela NC Mill in the Output Device option, and the .rml file extension appears in the Output File window. Don't forget to press ENTER as this tells CAM to accept the name, the toolset, and the parameters you've set thus far. Once you've done this, you'll have to reset x min and y min to 1 again, as CAM won't retain those numbers when you choose a new machine.

In the next row of buttons, you'll see Contour Boundary, Raster Interior, and Write Toolpath. If we click Contour Boundary, the program outlines in red all the areas that we want to keep, and clicking this button is mandatory to mill the board. If we want to etch away all the unwanted copper and just leave the traces that are important for connecting the components, click Raster Interior.

Figure B-9 Cam.py loading a file

The second line of buttons and boxes on the top menu bar starts with x min and y min. These refer to the origin of the x-axis and the y-axis on the bed of the milling machine. So the point where x = 0 and y = 0 on the milling machine is in the lower-left corner. If you started milling with x = 0 and y = 0, your machine would start cutting right at the lower-left corner of the milling machine bed. As mentioned before, we want to start the board

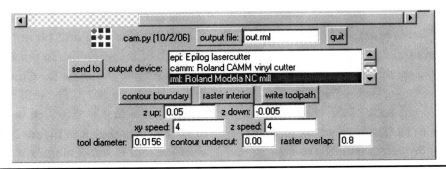

Figure B-10 More cam.py user controls

Using Write Toolpath, you can save the file with all of the parameters you have chosen. This allows you to open and mill the exact same file with the exact same dimensions, depth of cut, velocity, etc.

z up tells you how high above the material the end mill should lift to move from one place to another. The default setting is 0.05 inches. z down tells you how deeply into the material to cut. The default setting is –0.005 inches. xy speed tells you how fast the mill runs in the x and y directions. z speed tells you how fast the mill runs in the z direction. The default settings are 4 for both of these cases. Tool Diameter refers to the setting for the size of your end mill. The default setting matches with the 1/64-inch drill bit that we use for milling the tracks. Contour Undercut is a setting that controls where the center of the tool will cut in relation to the lines in your design object. It's a way to let you control precisely whether the tool cuts on the line, just inside the line, or just outside the line. If not needed, leave it to 0.0. Raster Overlap determines the resolution with which unwanted copper is etched away.

Once all these settings are in place, you are ready to send your board file to the machine for fabrication. The Send To button doesn't operate in Windows, as cam.py was designed for Linux-based systems and is coded so as to access the relevant serial port. The output .rml file generated can be sent to the machine using bray's terminal. This is described later.

Step 5: Milling, Drilling, and Cutting the PCB

Take two pieces of flame resistant 2 (FR2) copper board stock (one sacrificial and one useful), and turn one of them (the sacrificial one) upside down so that the copper is facing down. Put pieces of double-sided tape on the back of the copper board as shown in Figure B-11. The area where you will be milling needs to be solidly attached to the

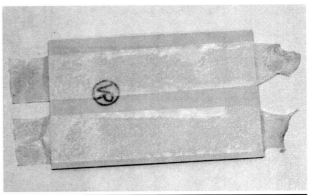

Figure B-11   Bare FR2 board with double-sided tape

surface of the Modela bed, so be sure to put lots of tape, close together, on the back, but not overlapping.

The material bed on the Modela has a grid of centimeter-sized squares. Place your material about two squares over and two squares up to accommodate your default (1,1) offset settings (basically, 1-inch offset is too large, and if we place our board leaving 2 cm, we give the net offset of 2.54 cm (1 inch) – 2 cm = 0.54 cm, which is good enough). After placing the sacrificial board on the bed, place the useful one over the sacrificial one in the similar way. Now using a piece of cloth or your shirtsleeve, firmly adhere the useful board over the sacrificial one by pressing down and rubbing back and forth over the copper board, as shown in Figure B-12. Don't do this with your fingers, as the oil from your skin can affect the conductivity of the copper traces. The sacrificial board is used to prevent the Modela bed from getting damaged when you drill the holes.

The next task is to set the drill bit into the correct position. The drill bit rests in a position in the back of the machine. To put it in active mode, press the View button on the Control Panel as shown in Figure B-13. The drill bit will move to its origin of x = 0, y = 0 (which is the lower-left side of the bed), as shown in Figure B-14. Now you want the mill to move to the x = 1, y = 1 position.

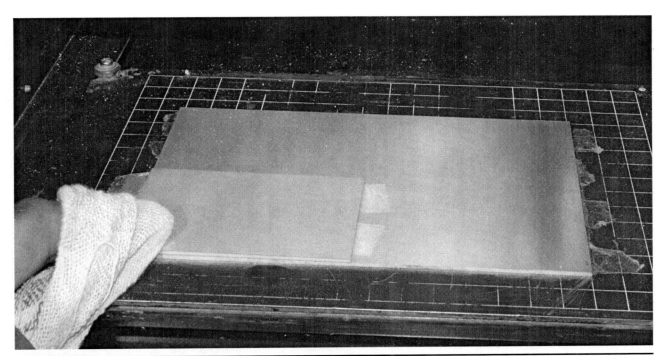

Figure B-12 Placing the bare board on the machine's bed

To do this, you need to send the commands to the machine by serial port. We use bray's terminal for this, which can be downloaded from www.avrgenius.com/tinyavr1. Open the terminal and configure the settings as shown in Figure B-15.

Figure B-13 View button

Check that the serial port name on your PC is com1. It may vary on your PC, however, so change that setting accordingly. After this, click the Connect button. Now you want the drill bit to move to the (1,1) position. For that, there is a file provided at www.avrgenius.com/tinyavr1 with the name move%d%d.txt. Open it, and it will show the following contents:

```
PA;PA;!PZ0,400;VS10;!VZ10;!MC0;PU%d,%d;
   !MC0;
```

In this text, replace %d,%d with the x- and y-coordinates of the position you want your drill bit to move to but pay attention in this file, because you have *to specify the coordinates in milli-inches instead of inches.* So to move the drill bit to (1,1) make the following changes:

```
PA;PA;!PZ0,400;VS10;!VZ10;!MC0;PU1000,
   1000;!MC0;
```

Copy this text in the Transmit section of the terminal (which is gray) and press ENTER. You will

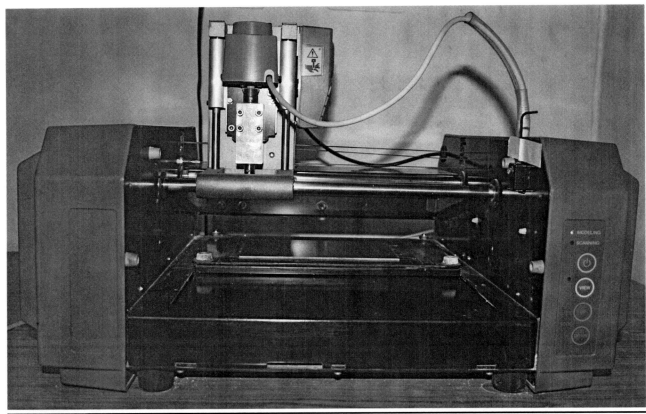

Figure B-14 Drill bit in origin (x = 0, y = 0) position

see the drill bit move to the (1,1) position over the copper board. Now back to the machine. We have to set the drill bit into position now. For milling, use a 1/16-inch drill bit, and for drilling and cutting, use a 1/32-inch drill bit. The drill bit can be held in place by a collet with two inset screws on opposite sides of the collet. Take the tiny Allen wrench that comes with the Modela, as shown in Figure B-16, and loosen the inset screws. Hold on to the drill bit firmly so you don't drop it and break the fine, fragile end, and push the drill bit fairly high in the collet. Tighten the screws slightly so that the drill bit is held in place, but not too

firmly. This is merely to get the drill bit safely out of the way while you position the carriage.

Your next move is to bring the entire drill bit carriage down to its lowest point—where the metal sides meet the metal base for the motor carriage. Do this by pushing and holding down the Down button on the Control Panel. Now you want to back off this position ever so slightly. So push and hold the Up button until the carriage moves up a little bit. What you are doing right now is indicating to the machine where Z = 0 is located. If you don't back off from the lowest position, the

	COM Port	Baud rate			Data bits	Parity	Stop bits	Handshaking
Connect	● 1 ○ 6	○ 600	○ 14400	○ 57600	○ 5	● none	● 1	○ none
ReScan	○ 2 ○ 7	○ 1200	○ 19200	○ 115200	○ 6	○ odd		● RTS/CTS
Help	○ 3 ○ 8	○ 2400	○ 28800	○ 128000		○ even	○ 1.5	○ XON/XOFF
About..	○ 4 ○ 9	○ 4800	○ 38400	○ 256000	○ 7	○ mark		○ RTS/CTS+XON/XOFF
Quit	○ 5 ○ 10	● 9600	○ 56000	○ custom	● 8	○ space	○ 2	○ RTS on TX

Figure B-15 Bray's terminal

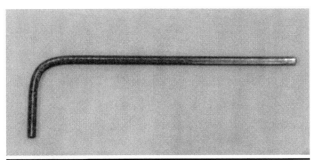

Figure B-16 Working with the Allen wrench

tool can't drill into the material. Now again, hold on to the drill bit with your finger, loosen the inset screws, and carefully place the drill bit on the surface of the copper, as shown in Figures B-17 and B-18. Now tighten the screws finger-tight such that the drill bit is firmly held in place.

Now you're ready to cut your circuit board. As a safety precaution, lift the drill bit off of the surface just above where you are going to start cutting, that is, (1,1).

To do that, go back to the computer terminal screen and type the following text again:

```
PA;PA;!PZ0,400;VS10;!VZ10;!MC0;PU1000,
    1000;!MC0;
```

Now you are ready to send the .rml file to mill the tracks. Select the Send File option on the terminal, and send the appropriate file. After milling is completed, remove the 1/16-inch drill bit and use the 1/32-inch drill bit. Repeat the procedure described earlier and send the .drillbot and .millbot files. After this, your board is ready to be soldered and tested.

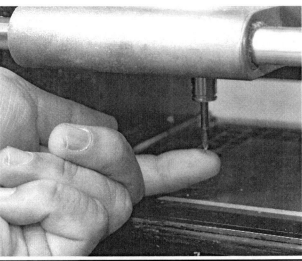

Figure B-17 Careful setting of the machine

Figure B-18 Final position of the drill bit

APPENDIX C

Illuminated LED Eye Loupe

IN CHAPTER 1, WE LOOKED AT a set of tools that would be useful for building and prototyping the various projects described in this book. One of the tools was the eye loupe, shown in the following illustration.

It's commonly used by watchmakers, and we find it very useful for inspecting components, printed circuit boards, etc. The photograph shows an eye loupe with 10× magnification. The problem with using an eye loupe in inspecting tiny components is the lack of illumination. In this appendix, we show how the eye loupe can be modified by adding white LEDs to provide illumination in the field of view. We actually show

three versions of the illuminated LED eye loupe. The first version uses a 9V battery and provides a fixed level of illumination; the second uses a single 1.5V AAA battery, and also provides fixed illumination; while the third version is a microcontroller-based solution that allows the user to adjust the illumination as per their requirements. The third version requires a 9V battery for operation.

All three versions require the eye loupe to be fitted with eight white LEDs with series current-limiting resistors as shown in the circuit diagram in Figure C-1. The LEDs have a 3-mm diameter. All eight LEDs with independent series resistors are connected in parallel.

To solder the LEDs and resistors, we need a piece of general-purpose printed circuit board (PCB) as seen here.

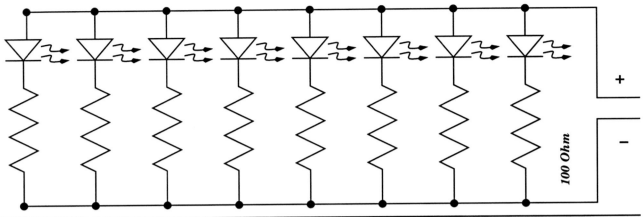

Figure C-1   LEDs and resistor circuit for the illuminated LED eye loupe

The PCB is cut in an annular shape, as shown in the next photograph, such that the lens end of the eye loupe fits snugly into the PCB.

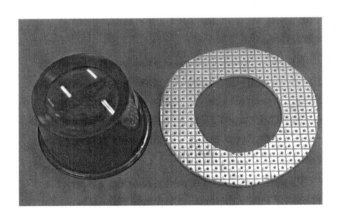

Once the PCB is cut, the LEDs are soldered uniformly around the hole. Please note that the LEDs are inserted from the solder side (the copper side) of the PCB and then soldered. Normally, components are inserted from the component side and soldered on the solder side, but here we have inserted the LEDs from the solder side. For soldering the LEDs, allow about half an inch of LED pins above the surface of the PCB. After soldering the LEDs, the LEDs are bent a little bit towards the center of the PCB.

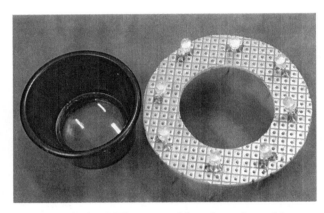

Once all the LEDs are soldered on the solder side of the PCB, the series resistors are installed from the component side of the PCB and soldered to individual LEDs, as shown in the following photograph.

Once the resistors are soldered to the LEDs, the other free ends of the resistors are soldered together. Also, the anodes of all the LEDs are soldered together, as seen in the following photograph.

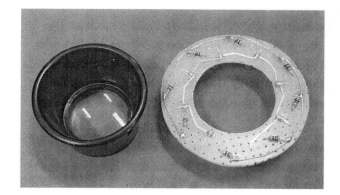

In the next step, two wires are soldered—one to the anodes of the LEDs and one to the shorted end of all the resistors. The LEDs are covered with hot glue to protect them from physical damage, as seen in the following photograph.

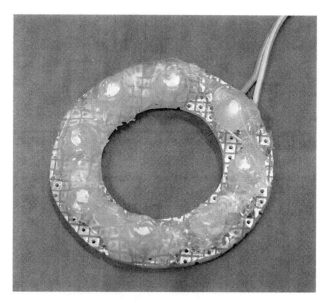

Once the wires are soldered to the PCB, they are connected to a 9V battery with an on/off switch in series so that you can turn on/off the LED when required.

The PCB assembly is now complete, and it's time to attach the eye loupe in the center of the PCB, as seen in the following two photographs.

Version 2 of the Illuminated LED Eye Loupe

The previous version of the illuminated LED eye loupe uses a 9V battery for powering the LEDs. However, a 9V battery is bulky and expensive. Using a 1.5V battery is preferred, but white LEDs need more than 3.5V for operation. So if operation with a 1.5V battery is desired, an electronic circuit to boost the battery voltage from 1.5V to, say, 4V is required. Such a voltage boost can be achieved

easily with a boost type DC-DC converter, but the added penalty is the high cost of a converter. A simple voltage boost circuit can be built with a single transistor oscillator called the relaxation oscillator, shown in Figure C-2.

It uses just three components: an NPN transistor, a special inductor, and a resistor. The output of such a relaxation oscillator is shown in Figure C-3, and it shows a pulsed waveform with more than 10V. This pulsed waveform is quite suitable to drive the white LEDs in our eye loupe.

The critical component of the circuit is the special inductor. It uses two coils of 36-gauge copper enameled wire wound together on a suitable former such as a ferrite bead or a toroid or a dumbbell. The dots on the two coils in Figure C-2

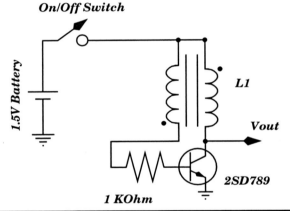

Figure C-2 Relaxation oscillator circuit diagram

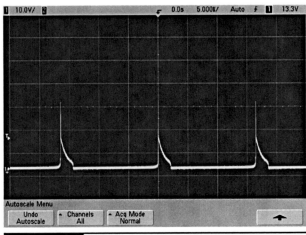

Figure C-3 Output of the relaxation oscillator

show the phase of the inductors. To build this inductor, start with enough length to wind about 10 to 20 turns. Take two equal lengths, as shown in the following illustration.

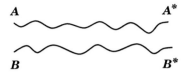

Twist them together. Also ensure that the insulation is removed from the ends (A and A* and B and B*) of the wires.

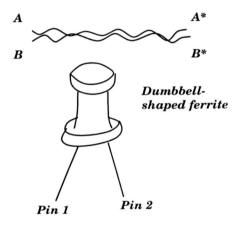

Dumbbell-shaped ferrite

Wind the twisted wires around the ferrite material (toroid, dumbbell, etc.) that you have chosen. The following illustration shows the wires

wound on a dumbbell-shaped toroid former with two pins.

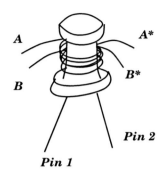

Solder any two opposite ends of different wires, either A and B* or B and A* together. Thus, from four wire ends you are left with three wire ends. These three wire ends are used for the inductor, as shown in Figure C-2.

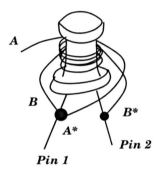

The soldered and completed circuit board is shown in the following photograph. The connector

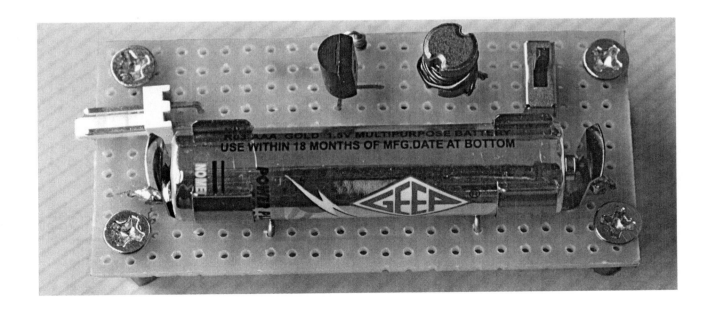

on the left is connected to the LED-and-resistor arrangement of the eye loupe.

Version 3 of the Illuminated LED Eye Loupe

The previous two versions of the illuminated LED eye loupe provide fixed light intensity. Sometimes, however, you may feel that the light is too bright, and at other times you might want some extra intensity. This version of the eye loupe will meet those needs. The project uses an eight-pin tinyAVR microcontroller. Tiny13 is quite suitable and sufficient for this project, although a Tiny24 or Tiny25 may also be used. Figure C-4 shows the circuit diagram. The circuit is powered with a 9V battery and uses the LP2950-5V voltage regulator to power the microcontroller. The circuit has a

potentiometer to set the desired intensity on the LEDs of the eye loupe. The potentiometer setting is read by the ADC channel of the microcontroller (pin 2) and is translated into a corresponding PWM signal on an output pin of the microcontroller (pin 7). This pin drives a medium-power NPN transistor. The collector of the transistor is connected to the −ve pin of the LED circuit (see Figure C-1). The positive terminal of the LED circuit is connected to the 9V battery of the circuit. The PWM signal varies the intensity from 100% to 0%, depending upon the setting of the potentiometer.

The following photographs show the soldered and completed circuit. The software for the controller is identical to the controller used in Project 3 (RGB LED color mixer) in Chapter 2.

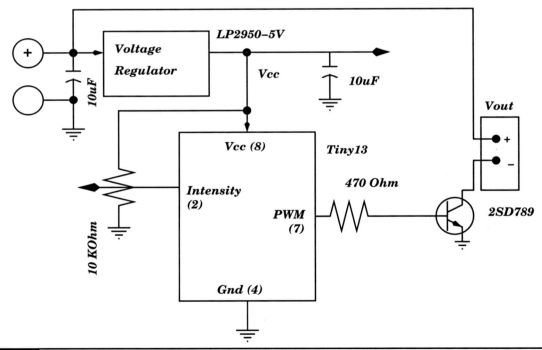

Figure C-4 Circuit diagram of the microcontroller-based LED intensity control circuit for the illuminated LED eye loupe

Index

References to figures are in italics.

A

AC adapters, 14–15
address counter, 102
alkaline batteries, 12
analog comparator, 6
analog to digital converter, 6
ANSI C, vs. embedded C, 23–24, 214
architecture, 3–4
    analog comparator, 6
    analog to digital converter, 6
    clock options, 6–7
    debugWIRE on-chip debug system, 8
    interrupts, 5
    I/O ports, 4
    memory, 4
    memory programming, 8
    power management and sleep modes, 7
    system reset, 7
    timers, 5
    universal serial interface (USI), 5–6
arrays, 222
assignment operator, 216
asynchronous clock sources, 5
ATtiny13, 2
ATtiny25/45/85, 2
ATtiny261/461/861, 2
ATtiny48/88, 2
audio feedback, 169–171
    fridge alarm redux, 176–178
    musical toy, 185–189
    RTTTL player, 178–185
    tone player, 171–176
autoranging frequency counter, 82–84
AVR Studio, getting started on a project with, 21–22

B

batteries, 11–13
    fruit battery, 14
batteryless electronic dice, 201–206
batteryless infrared remote, 196–201
batteryless persistence-of-vision toy, 206–211
bench vice, 20
bit banging, 101
bitwise operators, 217, 218–220
brownout reset, 7

C

C language, 20
    ANSI C vs. embedded C, 23–24, 214
    constants, 215–216
    data types, 214–215
    enumerated data types, 224
    floating point types, 215
    functions, 220–221
    operators, 216–217
C preprocessor, 223–224
C programming
    arrays, 222
    efficient management of I/O ports, 217–220
    header files, 220
    interrupt handling, 221–222
    overview, 213
C utilities, 222–224
calibrated resistor capacitor (RC) oscillator, 6
CAM.py, 232
Celsius and Fahrenheit thermometer, 80–82
Charlieplexing, 2, 65–67
    vs. multiplexing LEDs, 65
CLK_ADC, 6

CLK_CPU, 6
CLK_FLASH, 6
CLK_I/O, 6
clock options, 6–7
Colpitts oscillator, 153, 154
const qualifier, 224
contactless tachometer, 149–153
Conway, John, 113
copper braid, 17, *18*
crystal oscillator, 6

D

data memory space, 4
DDRAM, 103
debugWIRE on-chip debug system, 8
decoders, 62–63
delay.h, 220
devices, 2–3, *4*
DIP packaging, *2*
display data RAM. *See* DDRAM
drill machine, 19

E

EAGLE Light Edition, 225
    adding new libraries, 227–228
    placing the components and routing, 228
    tutorial, 226–227
    windows, 225–226
EEPROM memory, 4
electronic birthday blowout candles, 159–164
electronic dice, 201–206
electronic fire-free matchstick, 140–143
embedded computers, 1
enumerated data types, 224
external clock, 5, 6
external reset, 7
eye loupe, 17, *18*
    illuminated LED eye loupes, 239–245

F

Faraday-based generators, 16, *17*, 191–192
    building the Faraday generator, 194–195
    experimental results and discussion, 195–196
Faraday's law, 191
Fibonacci LFSR, 46
filter capacitor circuits, 14–15
flickering LED candle, 35–41
frequency counter, 82–84
fridge alarm, 164–167
fridge alarm redux, 176–178

fruit battery, 14
functions, 220–221

G

Galois LFSR, 36, 40–41
Game of Life, 113–116
geek clock, 84–90
GLCDs
    glitches, 104–105
    Nokia 3310, 100–104
graphical LCDs. *See* GLCDs

H

hardware development tools, 17–20
H-bridge, 171, 172, 176
header files, 220
high-voltage serial programming (HVSP), 8

I

illuminated LED eye loupes, 239–245
#include, 214, 223
inductive loop-based car detector and counter, 153–159
inductors, as magnetic field sensors, 131
infrared remote control devices, 196–201
in-system programming (ISP), 8
interrupt service routine (ISR), 5
interrupt.h, 220
interrupts, 5
    handling, 221–222
I/O ports, 4
    efficient management of, 217–220
io.h, 220

L

LCDs
    Game of Life, 113–116
    glitches, 104–105
    overview, 99
    principle of operation, 99–100
    rise and shine bell, 123–128
    temperature plotter, 105–109
    Tengu on graphics display, 109–113
    tic-tac-toe, 117–119
    Zany Clock, 119–123
LDRs, 130
LED displays, 2
LED pen, 49–54
LEDs
    autoranging frequency counter, 82–84

batteryless electronic dice, 201–206
batteryless persistence-of-vision toy, 206–211
Celsius and Fahrenheit thermometer, 80–82
Charlieplexing, 65–67
color and typical electrical and optical
    characteristics, 30
controlling, 32–35
electronic birthday blowout candles, 159–164
flicker, 33
flickering LED candle, 35–41
geek clock, 84–90
LED pen, 49–54
mood lamp, 67–72
multiplexing, 55–65
overview, 29–31
random color and music generator, 45–49
reverse-bias, 131
RGB dice, 90–93
RGB LED color mixer, 41–45
RGB tic-tac-toe, 93–97
as a sensor and indicator, 131–136
as sensors, 129–130
spinning LED top with message display, 144–149
types of, 31–32
valentine's heart LED display with proximity
    sensor, 136–139
voltmeter, 76–80
VU meter with 20 LEDs, 72–76
LFSR, 36, 40–41, 45–46
light doodles, 49–54
light-dependent resistors. *See* LDRs
light-emitting diodes. *See* LEDs
liquid crystal displays. *See* LCDs
lithium batteries, 12
logical operator, 216–217
low-frequency crystal oscillator, 6

M

M3 nuts and bolts, 19
macro substitution, 223
macros
    for AVR, 224
    vs. functions, 224
magnetic flux, 191
Make All, 20
Make Clean, 20
Make Program, 20
MAKEFILE Template, 20–21
mathematical operators, 216
memory, 4

memory programming, 8
mood lamp, 67–72
Moore, Gordon, 1
Moore's Law, 1
multimeter, 19
multiplexing LEDs, 55–65
    vs. Charlieplexing, 65
musical toy, 185–189

N

needle-nose pliers, 19
negative temperature coefficient (NTC), 130
nippers, 19
Nokia 3310, 100–104

O

operators, 216–217

P

PCBs, 14, 225
    fabricating, 228–237
    making your own, 24–26
    See also EAGLE Light Edition; Roland Modela
        MDX-20 PCB milling machine
PCD8455, 101–103
pgmspace.h, 220
phase lock loop (PLL) clock, 5
phase lock loop (PLL) oscillator, 6
picoPower technology AVR microcontroller class, 3
positive temperature coefficient (PTC), 130
POV toy, 206–211
power management, 7
power sources
    AC adapters, 14–15
    batteries, 11–13
    Faraday-based generators, 16, *17*
    fruit battery, 14
    RF scavenging, 16–17
    solar power, 16
    USB, 15–16
power-on reset, 7
printed circuit boards. *See* PCBs
program memory space, 4
Programmer's Notepad, 20
projects
    autoranging frequency counter, 82–84
    batteryless electronic dice, 201–206
    batteryless infrared remote, 196–201
    batteryless persistence-of-vision toy, 206–211
    Celsius and Fahrenheit thermometer, 80–82

projects *(continued)*
  contactless tachometer, 149–153
  electronic birthday blowout candles, 159–164
  electronic fire-free matchstick, 140–143
  elements of, 8–11
  flickering LED candle, 35–41
  fridge alarm, 164–167
  fridge alarm redux, 176–178
  Game of Life, 113–116
  geek clock, 84–90
  Hello World of Microcontrollers, 26–28
  inductive loop-based car detector and counter,
    153–159
  LED as a sensor and indicator, 131–136
  LED pen, 49–54
  mood lamp, 67–72
  musical toy, 185–189
  random color and music generator, 45–49
  RGB dice, 90–93
  RGB LED color mixer, 41–45
  RGB tic-tac-toe, 93–97
  rise and shine bell, 123–128
  RTTTL player, 178–185
  spinning LED top with message display, 144–149
  temperature plotter, 105–109
  Tengu on graphics display, 109–113
  tic-tac-toe, 117–119
  tone player, 171–176
  valentine's heart LED display with proximity
    sensor, 136–139
  voltmeter, 76–80
  VU meter with 20 LEDs, 72–76
  Zany Clock, 119–123
pulse width modulation
  pulse width modulated (PWM) signal, 33–34
  software-generated PWM, 43

R

random color and music generator, 45–49
random numbers, generating, 188
rechargeable batteries, 12–13
rectifiers, 14–15
reflective LCDs, 99–100
relational operator, 217
RF scavenging, 16–17
RGB dice, 90–93
RGB LED color mixer, 41–45
RGB LEDs, 31–32
RGB tic-tac-toe, 93–97
rise and shine bell, 123–128

Ritchie, Dennis, 23
Roland Modela MDX-20 PCB milling machine,
  228–237
RTTTL player, 178–185

S

screwdriver set, 19
sensors
  contactless tachometer, 149–153
  electronic birthday blowout candles, 159–164
  electronic fire-free matchstick, 140–143
  fridge alarm, 164–167
  inductive loop-based car detector and counter,
    153–159
  inductors as magnetic field sensors, 131
  LED as a sensor and indicator, 131–136
  LEDs as, 129–130
  spinning LED top with message display, 144–149
  valentine's heart LED display with proximity
    sensor, 136–139
serial peripheral interface (SPI), 6
shake detector, 202
silver oxide batteries, 12
sleep modes, 7
small form factor (SFF) PCs, 1
SMD packaging, *4*
software development, 20–24
solar power, 16
solder iron, 17, *18*
solder wire, 17, *18*
spinning LED top with message display, 144–149
square waves, 170
SRAM, 2, 4
static random access memory. *See* SRAM
Steinhart-Hart equation, 80, 130
supercapacitors, 140, 141, 192, 195–196
synchronous clock sources, 5
system reset, 7

T

tachometers, 149–153
temperature plotter, 105–109
Tengu on graphics display, 109–113
thermistors, 130
thermocouple, 80
tic-tac-toe, 117–119
timers, 5
Tiny devices, 2–3, *4*
Tiny form factor computers, 1
tinyAVR devices, 2–3, *4*

tinyAVR microcontrollers, 2
tone player, 171–176
transmissive LCDs, 100
transreflective LCDs, 100
tweezers, 19
two wire interface (TWI), 6

U

universal serial interface (USI), 5–6
USB, 15
    pins of the USB mini- or microconnector, 16

V

valentine's heart LED display with proximity sensor,
    136–139

volatile qualifier, 224
voltage regulators, 15
    choosing the right one, 192–194
voltmeter, 76–80
VU meter with 20 LEDs, 72–76

W

watchdog oscillator, 6
watchdog reset, 7
WinAVR, 20–21
    getting started on a project with, 23

Z

Zany Clock, 119–123
zinc-carbon batteries, 12

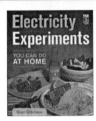

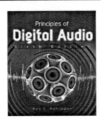

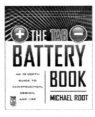

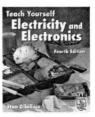

CPSIA information can be obtained at www.ICGtesting.com
Printed in the USA
BVOW04s1550211215

430741BV00009B/35/P